IMAGES
of America

Robert Goddard's Massachusetts

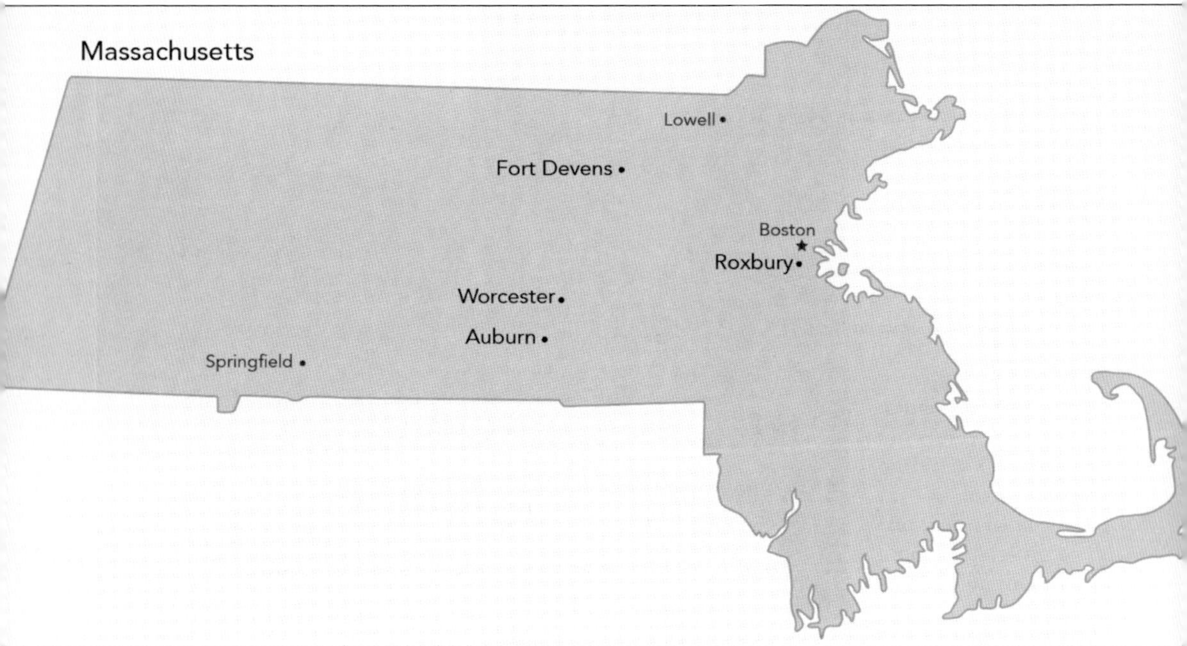

This map highlights four key Massachusetts locales—Worcester, Roxbury, Auburn, and Fort Devens—featured in this book, along with other notable cities across the state. Though Robert Goddard spent the majority of his life in Worcester, each of the other three locations played a vital role in shaping his development as a scientist and fueling the iconic innovations that defined his legacy. (Courtesy of Alex Elbert.)

On the Cover: On September 29, 1928, Robert Goddard stood by his "Hoopskirt Rocket," so nicknamed because of its resemblance to the once-fashionable hoop skirts worn by women. Goddard launched it a few months later, on December 26, 1928, marking his third successful launch of a liquid-fueled rocket. That flight covered a distance of 204.5 feet. Years later, Esther Goddard donated a reconstructed version of the rocket to the Smithsonian Institution's Air & Space Museum, a replica of which was created for display. (Courtesy of the Robert H. Goddard Library, Clark University.)

IMAGES
of America

ROBERT GODDARD'S MASSACHUSETTS

Kevin Schindler and Charles Slatkin
Foreword by Apollo astronaut Jim Lovell

ARCADIA
PUBLISHING

ISBN 978-1-4671-6314-9
Hardcover ISBN 978-1-5402-9954-3

Published by Arcadia Publishing
Charleston, South Carolina

Printed in the United States of America

Library of Congress Control Number: 2025945894

For all general information, please contact Arcadia Publishing:
Telephone 843-853-2070
Fax 843-853-0044
E-mail sales@arcadiapublishing.com

Visit us on the Internet at www.arcadiapublishing.com

To Robert and Esther Goddard and the Goddards of tomorrow, who dare to imagine the impossible and take the bold actions that transform vision into reality

Contents

Foreword

As someone who had the privilege of flying in space, I've long understood that our journey to the Moon and beyond did not begin with the countdown at Cape Kennedy. It began decades earlier, in quiet workshops and open fields, where a handful of visionary individuals laid the groundwork for space exploration. Among them, perhaps none was more influential than Dr. Robert H. Goddard.

During my high school years, against the backdrop of World War II, I learned about the German V-2 rocket—an engineering marvel that sparked my interest in rocketry. Around that time, I discovered a pamphlet by Dr. Goddard detailing his work on liquid-fueled rockets and reaching extreme altitudes. I was captivated, so much so that I built a model rocket that flew nearly 80 feet into the air.

That early fascination deepened at the US Naval Academy, where I wrote my term paper on liquid-fueled propulsion systems—an ambitious topic in the early 1950s. Goddard's work showed me that space travel wasn't just science fiction. Long before NASA existed, he believed that reaching the stars was not only possible—it was inevitable.

His vision, determination, and scientific rigor inspired me, as they did so many others, to pursue a career in aerospace. For me, Dr. Goddard was more than a pioneer—he was a guiding light whose ideas helped launch a generation of engineers, scientists, and astronauts.

This book, *Robert Goddard's Massachusetts*, captures the people, places, and moments that defined his early work. Through historic photographs and insightful captions, it brings to life the world in which he built his earliest rockets and proved that sustained, controlled rocket flight was achievable.

I am honored to offer this foreword and encourage readers to take time with each image. They tell the story of a man whose imagination helped lift us off this planet—and whose legacy continues to guide us toward the stars.

—Capt. James A. Lovell, US Navy (Ret.)
Apollo 8 and Apollo 13 Astronaut
July 30, 2025

Acknowledgments

The story of Robert Goddard is one of vision, inspiration, persistence, and the relentless pursuit of dreams. With that spirit in mind, we invited Capt. James Lovell to write the foreword for this book. Lovell, a lifelong admirer of Goddard's work, received the Goddard Trophy alongside his Apollo 8 crewmates for their groundbreaking contributions to aerospace. His own journey helped fulfill the dream Goddard had envisioned decades earlier—human spaceflight beyond Earth.

Captain Lovell's daughter Susan embraced our vision and played a vital role in helping her father craft the foreword after he graciously accepted our request. At 97 years old, Lovell remained deeply moved by the legacy of the man who had shaped his career. On July 30, Susan sent us the completed foreword. Just days later, to our profound sorrow, Captain Lovell passed away. As Susan shared, "It [writing the foreword] was the last thing he did here on this planet." We thus honor not only the pioneering rocket scientist Robert Goddard but also the legendary astronaut James Lovell and all the dreamers they continue to inspire.

Unless otherwise noted, images in this book are courtesy of the Archives and Special Collections at Clark University's Robert H. Goddard Library. Special thanks to Cindy Shenette for her tireless help in locating materials and to Laura Robinson, Clark's university librarian, for an unforgettable conversation and private viewing of an original Goddard rocket frame.

We also extend our gratitude to Gillian McCuistion and Emily O'Brien of Worcester Polytechnic Institute's George C. Gordon Library, Wendy Essery of the Museum of Worcester, and Ian McLennan for generously providing additional images.

Thanks to Alex Elbert for creating the map and to fellow Lowell Observatory marketing team members Cody Half-Moon, Sarah Gilbert, Alex Elbert, and Madi Mooney. We appreciate Mary DeMuth, Bill Munch, Danielle Adams, Iris Fisher, and Jeff Hall for reviewing early drafts. We're grateful for the inspiration and support of Alan Stern, David Eicher, Sara Putnam, and Colleen Dougherty. Thanks to Erin Vosgien and Caroline Anderson Vickerson at Arcadia Publishing, who helped bring this book to life.

Kevin Schindler wishes to thank Gretchen, Mom, Dick, Alicia, Addison, Sommer, Brandon, Senna, June, Lauren, Aaron, Donnie, Cherrie, Terry, Kim, Tony, and their families for ongoing inspiration. Charles Slatkin wants to thank his family and the countless volunteers and supporters of the Goddard project and its many initiatives to help inspire young and old.

A heartfelt thank you to the Goddards of today and those yet to come, whose stories, spirit, and generosity we have had the privilege to encounter along this journey. Your passion and curiosity continue to inspire us all to dream boldly and reach for the stars.

INTRODUCTION

Robert Goddard's name may not be as instantly recognizable as the Wright brothers, Charles Lindbergh, or Neil Armstrong, but his legacy is no less significant. He has been called "the father of the space age," "the father of American rocketry," and "the father of modern rocketry." If there were a Mount Rushmore for air and space pioneers, Goddard's face would surely be carved into it. A native of Massachusetts—a state steeped in innovation and bold ambition—Goddard revolutionized the future with his early-20th-century development of liquid-fueled rockets. His visionary work didn't just lay the groundwork for humanity's journey into space, it ignited the imaginations of the very pioneers who transformed his dreams into reality. In this book, we take advantage of a trove of pictures to share Goddard's scientific and technological accomplishments as well as introduce some of the people who inspired him, some of the people he himself inspired, the places he lived, and the legacy he left.

Born in 1882 in his family's multigenerational home in Worcester and passing in 1945, Goddard spent nearly 50 years living and working in Massachusetts, his eyes fixed on the stars. Driven by a blend of quintessential American values—vision, devotion, persistence, and inspiration—and buoyed by unwavering family support, he pursued and ultimately achieved his lifelong goal of developing the means for space travel.

Raised in Roxbury and Worcester, Goddard overcame childhood illnesses and family hardships, including his mother's battle with tuberculosis. He earned degrees from Worcester Polytechnic Institute and Clark University, laying the academic groundwork for his pioneering career. In 1912, he accepted a position at Princeton University, but within a year, tuberculosis forced him to return to Worcester. Though doctors expected him to succumb to the illness, Goddard defied the odds. His recovery is a testament to the raw determination that would define his life.

A voracious reader and early diarist, Goddard's sense of purpose crystallized during a transformative moment on October 19, 1899, when he was 17 years old. While trimming branches that evening in his favorite cherry tree behind his Worcester home, he experienced a dreamlike vision: developing a device capable of traveling into space. He climbed down from that tree forever changed, newly awakened to a mission that would guide him for the rest of his life. As he wrote in his *Material for an Autobiography*, "Life now had purpose for me."

After decades of planning and experimentation, Goddard launched the world's first liquid-fueled rocket on March 16, 1926, from the farm of his "Aunt Effie" Ward in Auburn, a handful of miles from Worcester. The flight lasted just 2.5 seconds and reached an altitude of 41 feet. Only five people witnessed it, and not even the local newspapers reported on the event. Yet its significance was historic; just as the Wright brothers' 1903 flight ushered in the age of aviation, Goddard's launch laid the foundation for space travel.

Goddard continued refining his designs and launching several more rockets from Aunt Effie's farm. But a 1929 launch sparked a fire in nearby fields, drawing media attention and prompting safety concerns. Ordered by a fire marshal to cease operations, Goddard relocated for a short time to Camp Devens, a US Army facility located 28 miles to the northeast of Goddard's Worcester home base.

Ironically, the same 1929 launch that forced his move also brought a stroke of luck. Legendary aviator Charles Lindbergh, fresh off his historic 1927 solo flight from New York to Paris, took notice. Impressed by Goddard's work, Lindbergh persuaded his wealthy friend Daniel Guggenheim to fund Goddard's rocket research. This financial lifeline allowed Goddard to move to the more arid climate of Roswell, New Mexico, where he expanded his test flight operation and continued to develop crucial supporting instruments such as gyroscopes and stabilizers.

In the final three years of his life, Goddard worked in Maryland developing rocket-assisted aircraft for the US Navy during World War II. He died of throat cancer on August 10, 1945, just weeks before the war's end.

Goddard's achievements extend far beyond his historic launch, and his innovations laid the groundwork for modern rocketry. Among his most notable accomplishments:

- Launched the first liquid-fueled rocket
- Experimentally proved that a rocket will provide thrust in a vacuum
- Patented multi-stage rockets
- Launched the first scientific payload on a rocket
- Invented gyroscopic stabilization for rockets
- Developed a handheld rocket launcher (precursor to the bazooka)
- Conducted groundbreaking research on solar energy and its potential as a power source

Goddard's lifework was truly revolutionary and inspirational, and he belongs among the immortals of air and space history. Yet he would have been the first to point out that he didn't achieve his accomplishments alone. His family nurtured his scientific interests, providing him with microscopes, telescopes, and unflagging encouragement. High school teachers and college professors, as well as the science fiction writings of H.G. Wells and others, encouraged and inspired him throughout his life.

Most vital to Goddard's success and legacy was his wife and life partner, Esther. She photographed and filmed his launches, transcribed his notes, sewed his rocket parachutes, and stood by him through triumphs and setbacks.

Esther outlived her husband by decades, and she spent the majority of the rest of her life tirelessly working to preserve his legacy. She transcribed and archived his diaries and notebooks, secured recognition for dozens of patents, won a $1-million lawsuit against the US government for unauthorized use of his inventions, and advocated for NASA to name one of its key research facilities—the Goddard Space Flight Center—after him. Simply put, without Esther, Robert Goddard would not have met the success he did, and his legacy would not be nearly as well documented.

Goddard's story is part of a larger cycle of inspiration. In 1877, Italian astronomer Giovanni Schiaparelli observed linear markings on Mars. This inspired Boston industrialist and Orientalist Percival Lowell to study the planet in hopes of proving the existence of intelligent life. The charismatic Lowell informed, if not inspired, writers such as H.G. Wells, who in 1898 penned the science fiction classic *War of the Worlds*. Goddard read this tale at an early age, and it helped inspire him to spend his life developing a means of traveling to space. Goddard's work in turn inspired future space explorers such as Jim Lovell, who helped turn Goddard's vision of space travel into reality.

This cycle of inspiration continues, as Goddard, Lovell, and their fellows inspire future generations of explorers to dream big and pursue their dreams with relentless determination.

Robert Goddard, the very embodiment of vision, devotion, persistence, and inspiration, successfully turned his dreams into reality. May his example continue to inspire dreamers around the world.

Robert Goddard's Milestones

1882 (October 5)	Born in Worcester in parents' multigenerational home
1883	The Goddard family moves to Roxbury
1898	Mother contracts tuberculosis, family moves back to Worcester
1899 (October 19)	Experiences cherry tree vision, calls October 19 "Anniversary Day"
1901 (March 31)	Esther Christine Kisk is born in Worcester
1904	Graduates as "best male student" at South High School
	Delivers "On Taking Things for Granted" speech at Mechanics Hall
1908	Earns bachelor of science degree (physics) from Worcester Polytechnic Institute
1910	Earns master of arts degree (physics) from Clark University
1911	Earns doctorate (physics) from Clark University
1912–1913	Works at Princeton University on a fellowship
1913	Contracts tuberculosis and moves back to Worcester
1914	Awarded first two patents for rocket apparatus
1916	Proves experimentally that a rocket will work in a vacuum
1917	Receives first grant: $5,000 from the Smithsonian Institution
1919	Publishes *A Method of Reaching Extreme Altitudes*
1924 (June 21)	Marries Esther Christine Kisk
1926 (March 16)	Launches first liquid-fueled rocket in Auburn
1929 (July 17)	Launches first rocket with science payload
1929 (November 23)	First meets Charles Lindbergh
1930–1942	Conducts research in Roswell, New Mexico, except from 1932 to 1934
1932	Invents gyroscopic stabilization for rockets
1942	Moves to Maryland
1942–1945	Works at the Naval Engineering Experiment Station
1945 (August 10)	Dies of throat cancer
1982 (June 4)	Esther dies

One

Roxbury

Youthful Curiosity

I imagine an innate interest in mechanical things was inherited from a number of ancestors who were machinists.

—Robert Goddard, *Material for an Autobiography*, 1927/1933

Robert Goddard was born in Worcester, but Roxbury, a historic neighborhood in Boston that had been originally settled in 1630 by the Massachusetts Bay Company, became the formative backdrop for his early life. In 1883, Nahum and Fannie Goddard moved from Worcester to Roxbury, bringing their infant son, Robert, with them. The family initially rented a home before purchasing a modest house at 63 Forest Street, nestled in a residential area. Nahum found work with a knife manufacturing company and later partnered with Simeon Stubbs to invest in a knife business of his own.

Robert spent most of his childhood and early adolescence in Roxbury, with occasional visits to Worcester to reconnect with extended family and friends. His days were filled with activities of a typical late-19th-century childhood—attending school and church, performing in plays and musical programs, and playing in the streets with neighborhood companions.

Among Goddard's most cherished activities was reading, a passion that remained constant throughout his life. He frequently visited the Roxbury Public Library, where he immersed himself in books that fed his imagination and curiosity. He was also a devoted reader of the *Youth's Companion*, a popular magazine for children that offered stories and moral lessons. These readings sparked his interest in invention and exploration, laying the intellectual groundwork for his later scientific pursuits.

Family outings added further richness to his upbringing. The Goddards often visited nearby harbors to watch ships, stirring Robert's fascination with engineering and motion. They took historical excursions to landmarks like the Bunker Hill Monument, connecting him to the legacy of American innovation and persistence. Nature expeditions to Franklin Park, a sprawling green space designed by renowned landscape architect Frederick Law Olmsted, offered Robert a chance to observe the natural world up close.

It was in Roxbury that Robert began to cultivate his skills as a scientific observer and experimenter. Surrounded by books, nature, and a supportive family, he developed the habits of critical thinking and devotion to solving a problem that would later define his groundbreaking work in rocketry. Roxbury wasn't just where he grew up, it was where his mind took flight.

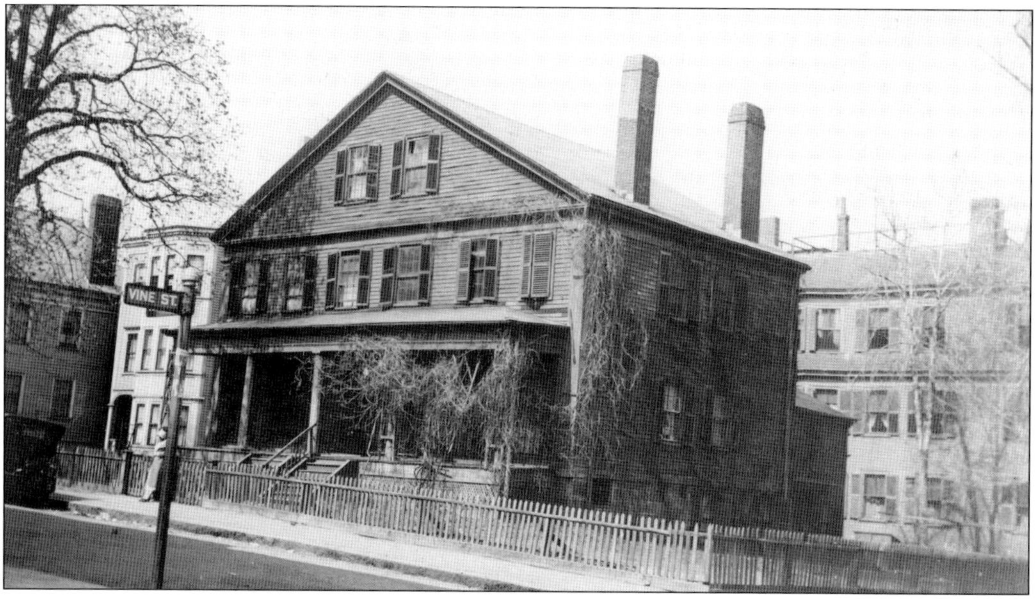

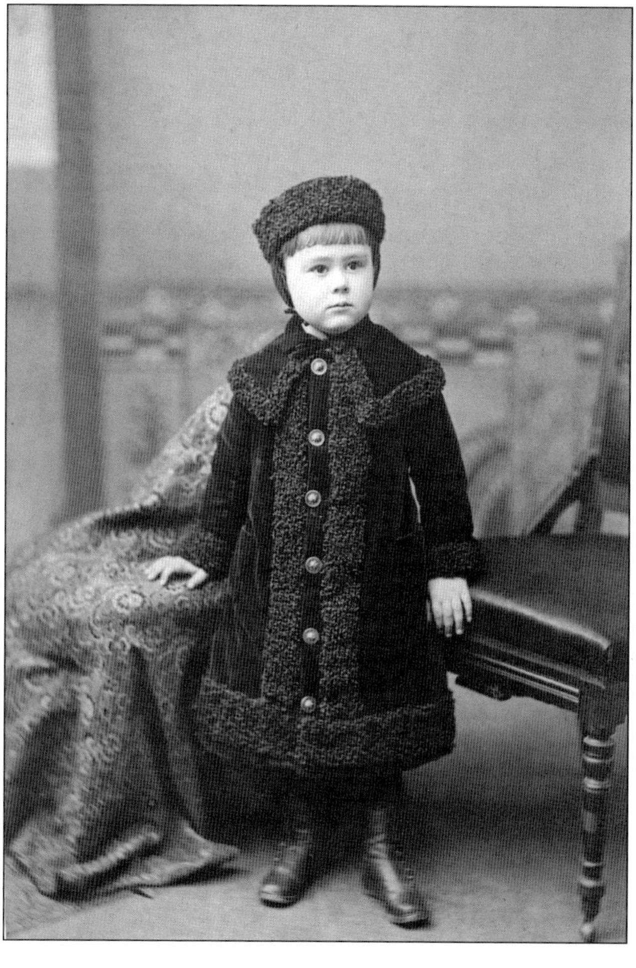

Within a year of Goddard's birth on October 5, 1882, his father, Nahum, relocated the family to the Boston suburb of Roxbury, where they lived for several years at 63 Forest Street. The house shown in this photograph is believed to have been their residence during that time. It served as a home base from which he began exploring the universe.

Goddard poses for a studio portrait at around the age of three, already showing signs of the curiosity that would define his life. Even at this young age, he seems to observe the world with the eyes of a budding scientist. His parents recalled that little Robert could be captivated for hours simply by sitting near a railroad yard, watching the cars move back and forth.

Goddard is shown here likely just weeks before his fifth birthday. It was about this time, he later recalled, that he conducted his first scientific experiment. Inspired by a budding curiosity about energy, he attempted to see if he could jump farther by scuffing his feet across the floor and using a battery to create an electric charge. The experiment marked the beginning of a lifelong fascination with physics and propulsion.

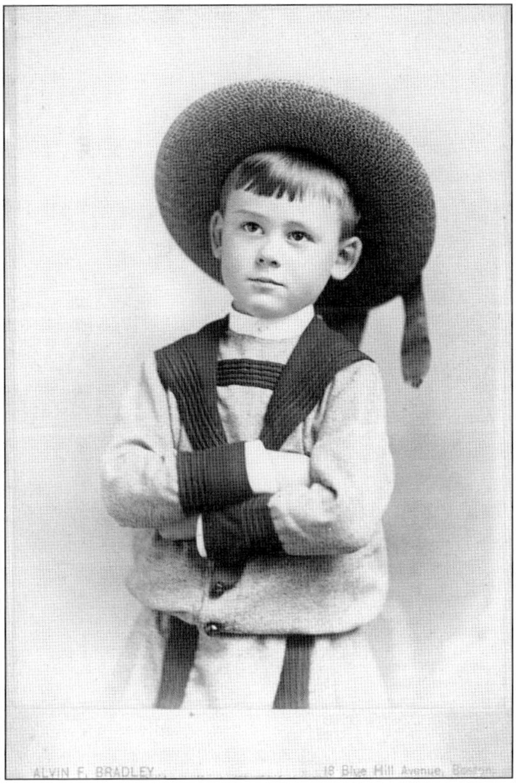

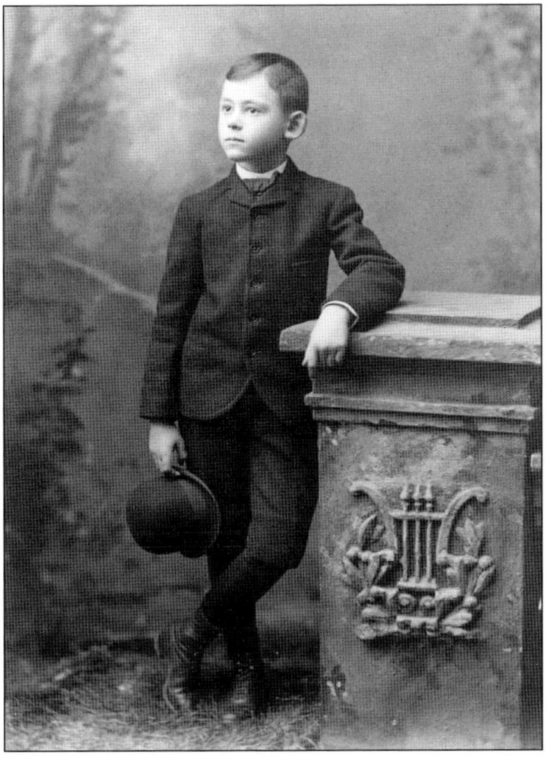

Shown here at the age of eight, Goddard for a time aspired to become a physician, inspired by the detailed illustrations of eyes and ears in a school physiology book. As he later recalled with a touch of humor, "I tested the eyesight of my great-grandmother Ward, who, being a paralytic, could not get away and whose feeble eyesight proved ideal for the purpose."

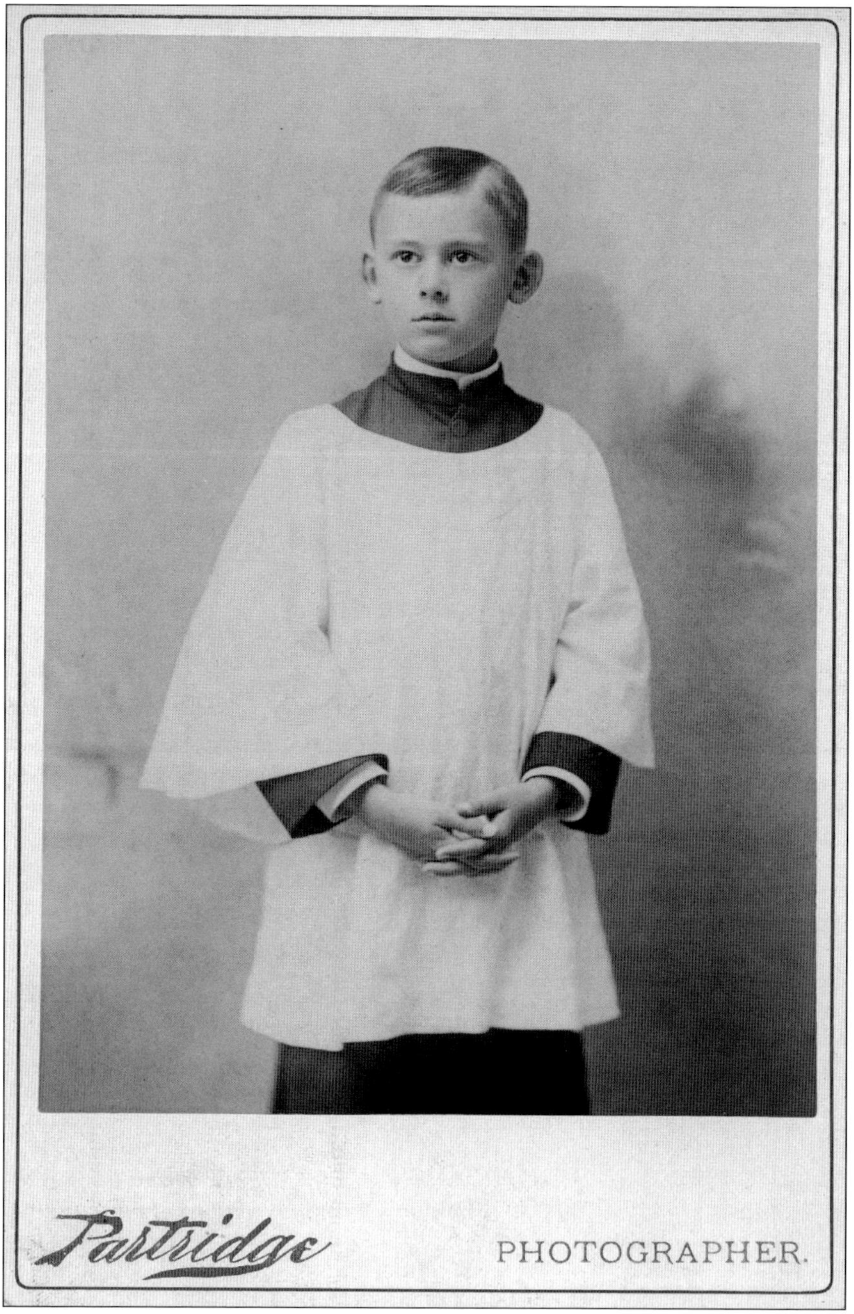

At about the time he was considering becoming a physician, Goddard was photographed in his altar boy attire, a reflection of his family's ties to church life in Roxbury. He regularly attended services and Sunday school, sang in the choir, and participated in pageants and holiday performances at St. Anne's Church. During one memorable event at nearby Winthrop Hall called "Sixteenth-Century Christmas," he was cast in the modest role of a street singer, but his costume told a different story. Instead of wearing rags, Goddard appeared in the elegant attire of a prince. The costume came from Caroline Goodwin, a neighbor whose son Nat Goodwin was a celebrated Boston stage actor. She had saved some of Nat's old theatrical clothes and gave them to Goddard.

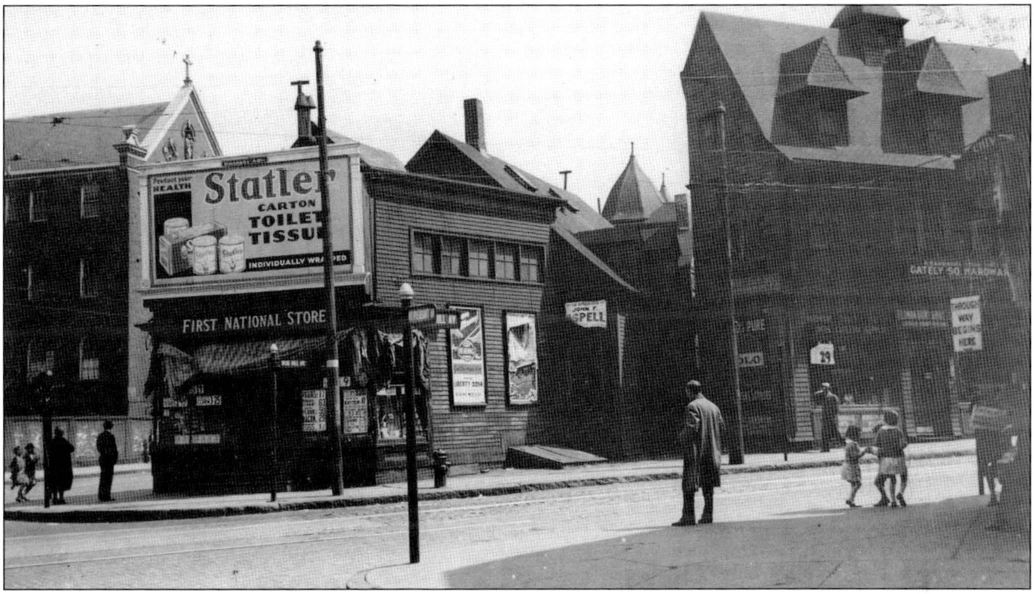

Although this c. 1928 photograph of Roxbury shows it decades after Goddard lived there, it still captures the essence of the neighborhood during his youth. One of his favorite places to visit, not shown in this photograph, was the Roxbury Public Library. There, he immersed himself in books that opened windows to worlds far beyond his own. The library became a sanctuary of imagination, nurturing his growing fascination with science.

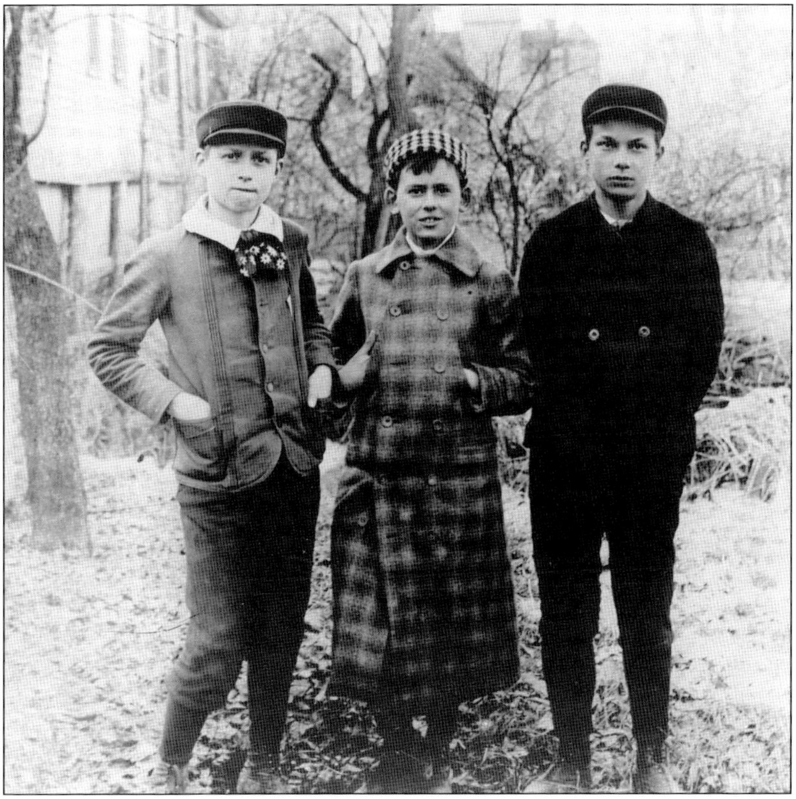

Some of Goddard's best friends in Roxbury were his neighbors, the Hemenways. Robert is here flanked by Russell B. (left) and F.M. Hemenway. The boys enjoyed a variety of adventures, from ice skating and tobogganing in the winter to beginning to dig a tunnel to China during one summer.

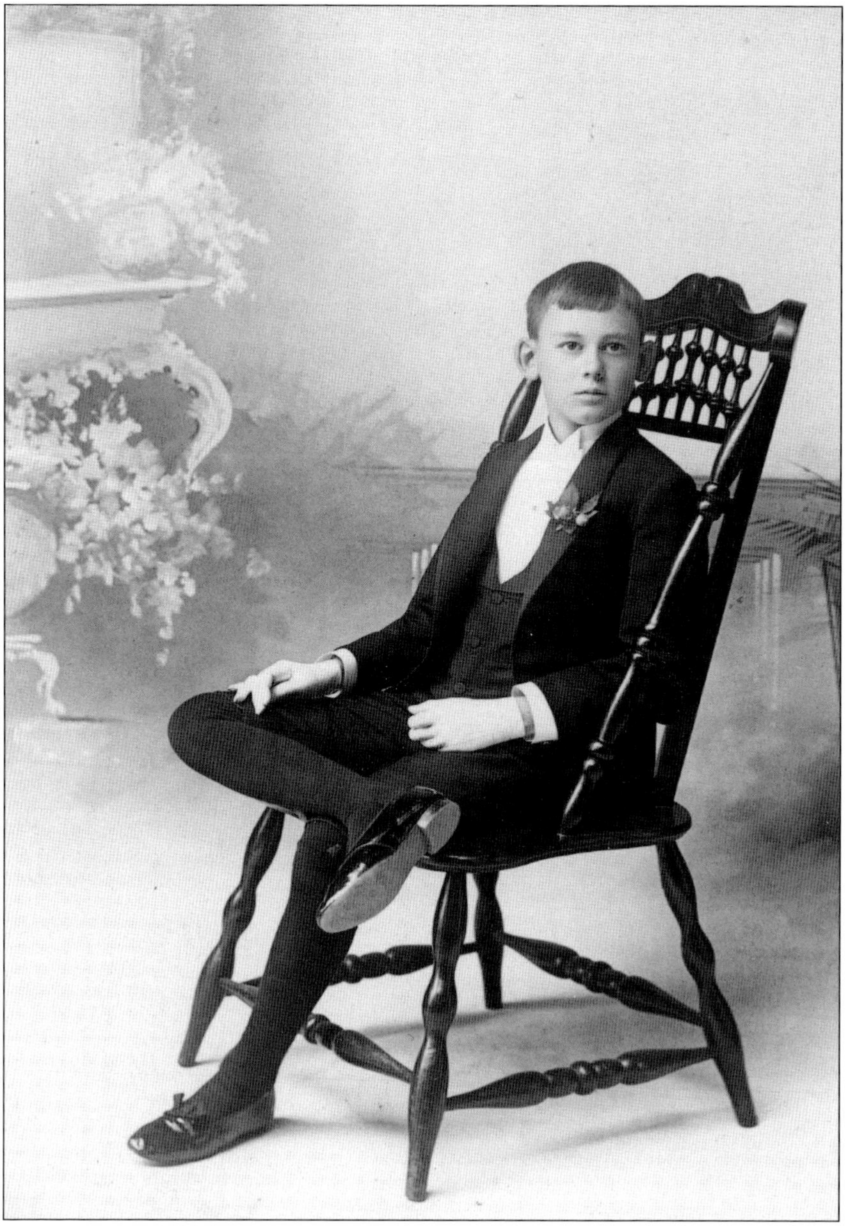

At the age of 10, Goddard balanced the ordinary routines of childhood, such as serving as a wedding attendant, with a growing love for the outdoors. While living in Roxbury, his family often visited a family friend in Worcester affectionately known as "Aunt Lizzie" Woodward. These trips exposed him to a world far removed from city streets. In Worcester's forests and ponds, Robert spent hours with his father fishing, taking photographs, and exploring nature. He was captivated by the quiet beauty of the outdoors and eagerly absorbed every detail. On one of these visits, he imagined building a frog hatchery, a project that reflected his early scientific curiosity. He researched frogs, designed a small one-by-three-foot structure, and prepared to experiment. Though he never used the hatchery, the dedication he showed foreshadowed the passion that would later drive his groundbreaking work in rocketry. Even as a child, Robert Goddard was not just observing nature—he was beginning to think like a scientist.

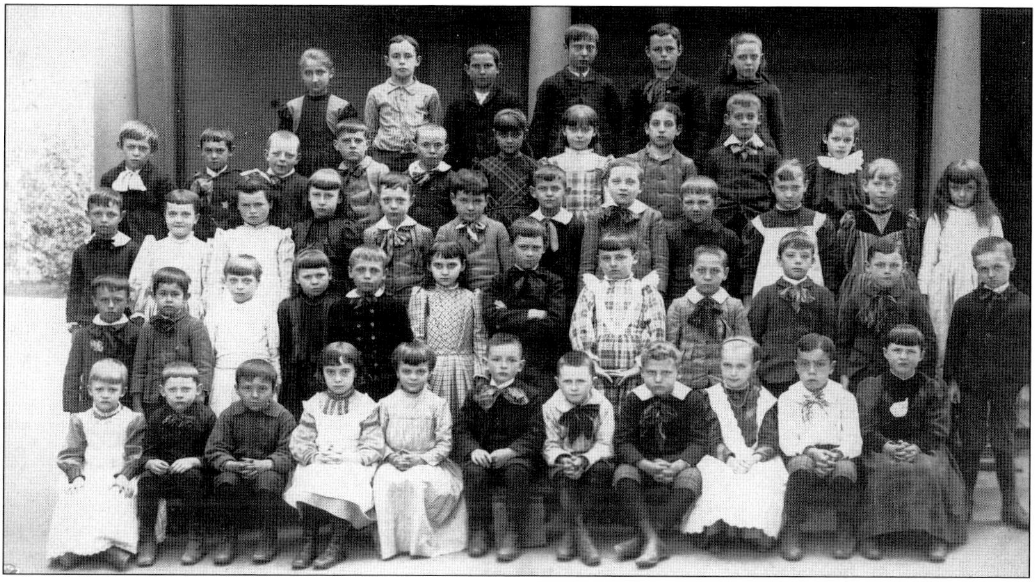

Goddard attended three schools while living in Roxbury. Here he is shown at the age of nine in a class photograph at Mount Pleasant Primary School; he is the one standing in the second row with his arms crossed. He later attended the newly opened Hugh O'Brien Elementary School, followed by a year at English High School before dropping out due to illness.

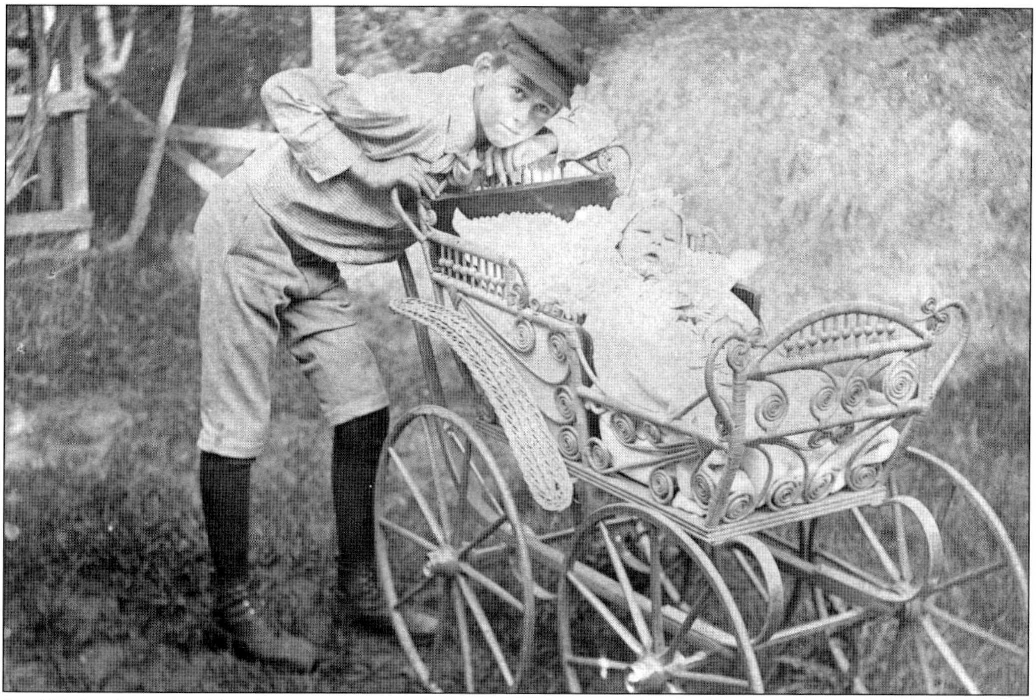

In 1894, when Goddard was 11 years old, his family welcomed a new addition with the birth of his brother Richard. Robert, shown here doting on Richard, embraced having a companion, but his joy was short lived. Richard died within the year due to a spinal deformity. The death cast a long shadow over the Goddard household. For Robert's parents, their sorrow translated into a heightened protectiveness toward their surviving son.

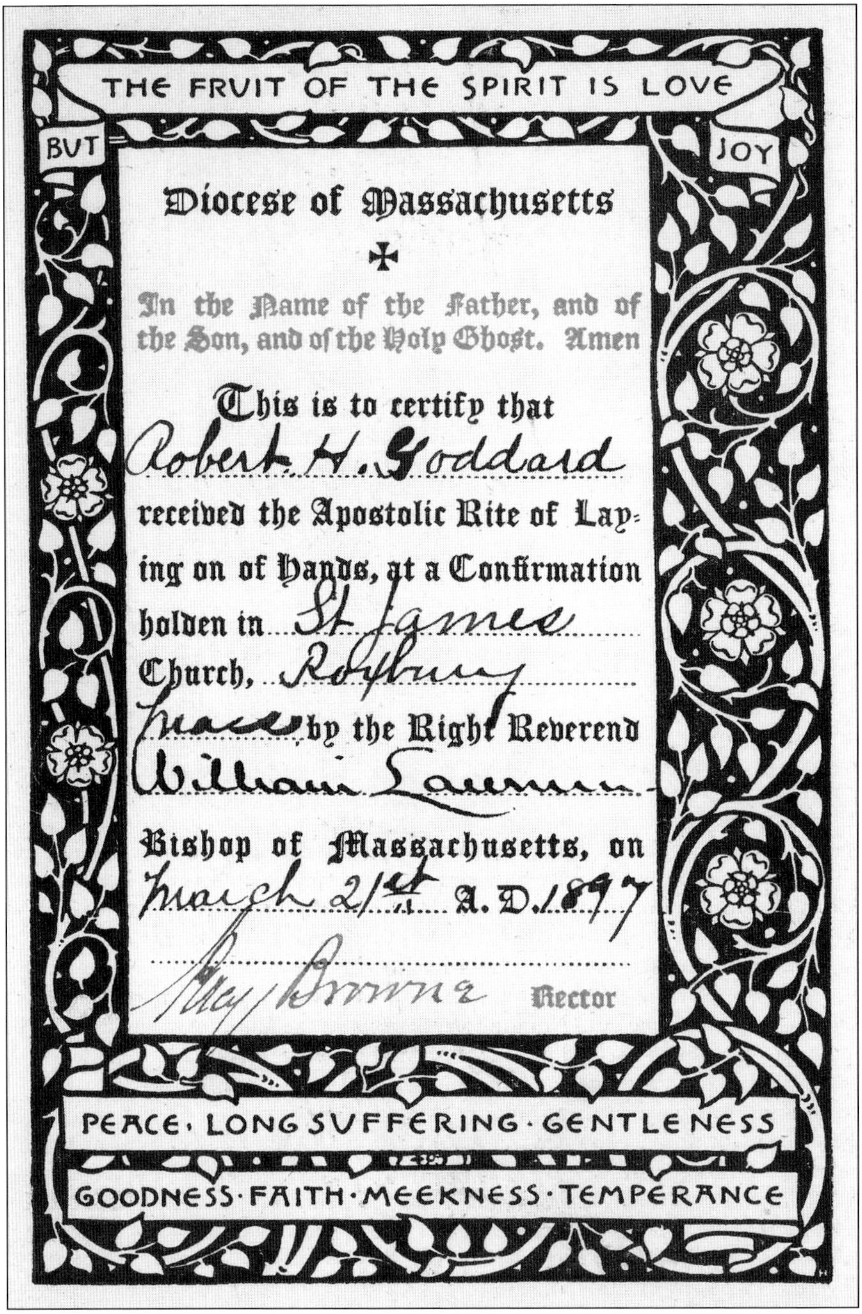

THE FRUIT OF THE SPIRIT IS LOVE

BUT JOY

Diocese of Massachusetts

✠

In the Name of the Father, and of the Son, and of the Holy Ghost. Amen

This is to certify that

Robert H. Goddard

received the Apostolic Rite of Laying on of Hands, at a Confirmation holden in St James Church, Roxbury Mass by the Right Reverend William Lawrence

Bishop of Massachusetts, on March 21st A. D. 1897

[signature] Rector

PEACE · LONG SUFFERING · GENTLENESS

GOODNESS · FAITH · MEEKNESS · TEMPERANCE

At the age of 14, Goddard received a certificate for participating in the religious rite of the laying on of hands, a central element of the Christian confirmation ceremony. This rite, practiced in many denominations such as Catholicism, Anglicanism, and Lutheranism, symbolizes the imparting of the Holy Spirit and marks a young person's deeper commitment to their faith. Typically performed by a bishop or elder, the laying on of hands is meant to strengthen the individual spiritually and affirm their place within the religious community. For Goddard, this moment may have held personal significance, as these early spiritual awakenings may very well have inspired, to some degree, his awe of the universe and his drive to explore it.

Two

WORCESTER

CRADLE OF MODERN SPACEFLIGHT

It has often proved true that the dream of yesterday is the hope of today and the reality of tomorrow.

—Robert H. Goddard, high school graduation speech, 1904
(as recorded in volume one of *The Papers of Robert H. Goddard*, 1970)

Worcester, Massachusetts, known as the "Heart of the Commonwealth" and the second-largest city in New England, was a hub of the American Industrial Revolution in the 19th century. Textile mills, machine shops, and wire production plants sprang up across the city, fueling a wave of technological innovation. In many ways, it served as the "Silicon Valley" of its time. Amid this industrious backdrop was born the future father of modern rocketry—Robert H. Goddard. Worcester remained a touchstone throughout his life.

Goddard was born in the Worcester home his family would own for more than 150 years, affectionately known as Maple Hill. Not long after his birth, the family relocated to Roxbury, where Robert spent much of his early childhood. Despite the move, he frequently returned to Worcester for family visits and outdoor excursions, including hunting and fishing trips.

After his mother contracted tuberculosis in 1898, the family moved back to Worcester. Robert would repeat this pattern of returning to the city throughout his life. After earning his doctorate from Clark University, he accepted a research position at Princeton University, but he soon contracted tuberculosis himself and returned to Worcester to recover. He lived there until relocating his rocket development efforts to Roswell, New Mexico, in 1930. Even then, Worcester continued to draw him back through the years.

Following his death in Maryland in 1945, Goddard returned once more—this time permanently—as he was laid to rest in the Goddard family plot at Hope Cemetery, just two miles from Maple Hill.

Many of Goddard's most pivotal moments occurred while living in Worcester: his visionary cherry tree dream of interplanetary travel; his graduation as the top male student from South High School in 1904; and the development of his groundbreaking liquid-fueled rocket, which he launched in nearby Auburn.

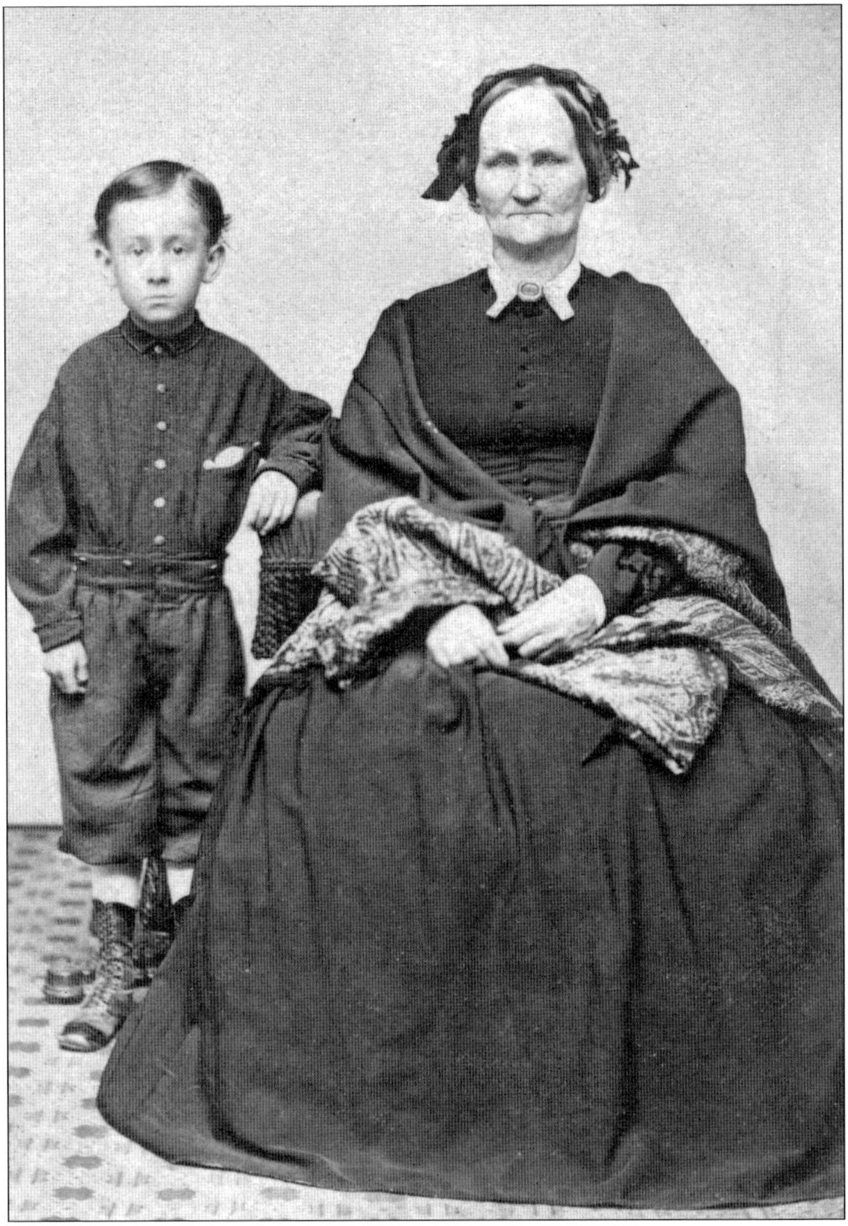

Robert Goddard's family roots in America stretch back centuries. His paternal ancestors arrived from England in 1666. Generations later came Nahum Parks Goddard. He would grow up to be a musician, marry the headstrong Mary Pease Upham, and father Nahum Danford Goddard, shown here with his mother. Not able to make a decent living in Boston, the family moved to Worcester in the 1870s to live at the house where Robert would later be born and spend much of his adult life. On Robert's maternal side, the family's American story began in 1683. Among his notable ancestors was Capt. Levi Pease, who operated the first stagecoach lines in the United States. After relocating to Worcester, Nahum Danford found work with the L. Hardy Company, owned by neighbor Henry Hoyt. He soon began courting Hoyt's daughter, Fannie. Despite the objections of Hoyt, who viewed the Goddards as socially inferior, Nahum and Fannie married in January 1882. Later that year, they welcomed their first child: Robert Hutchings Goddard.

Fannie Louise Hoyt was born in 1864 and died in 1920. Shown here in a picture taken a year before she married Nahum Danford Goddard, she grew up in a wealthy Worcester family before marrying him. She contracted tuberculosis in 1898 while she, her husband, and son Robert lived in Roxbury. The illness spurred them to move back to Worcester, where family and country air could help with her recovery.

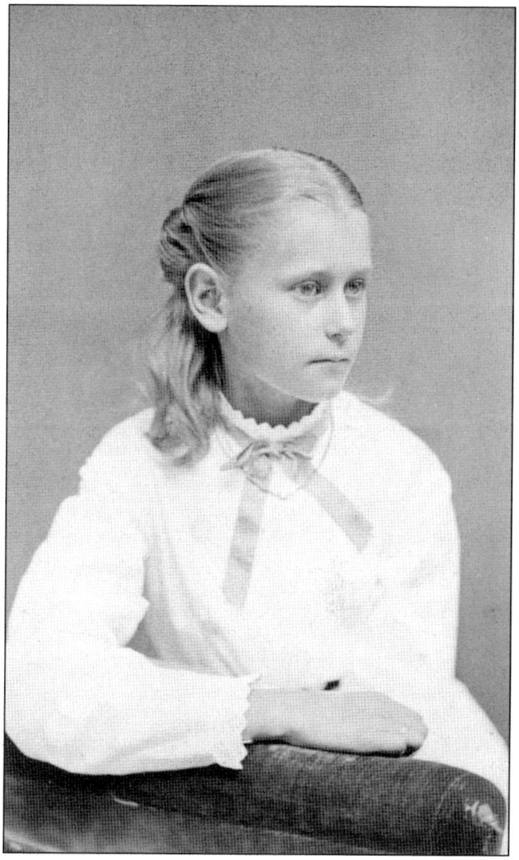

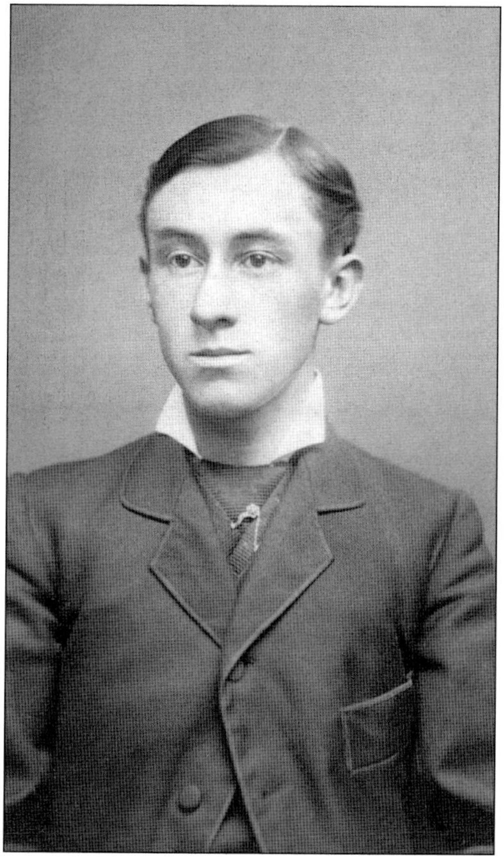

Nahum Danford Goddard, shown here about a year before his marriage to Fannie, was born in 1859 and died in 1928. A skilled inventor and salesman, Nahum built a career in the machine knife industry, where he developed and patented a range of specialized knives and mechanical devices. Beyond his professional life, he nurtured his son Robert's early interests in science and the natural world. After Fannie died in 1920, Nahum married widower Jennie Giles Ward.

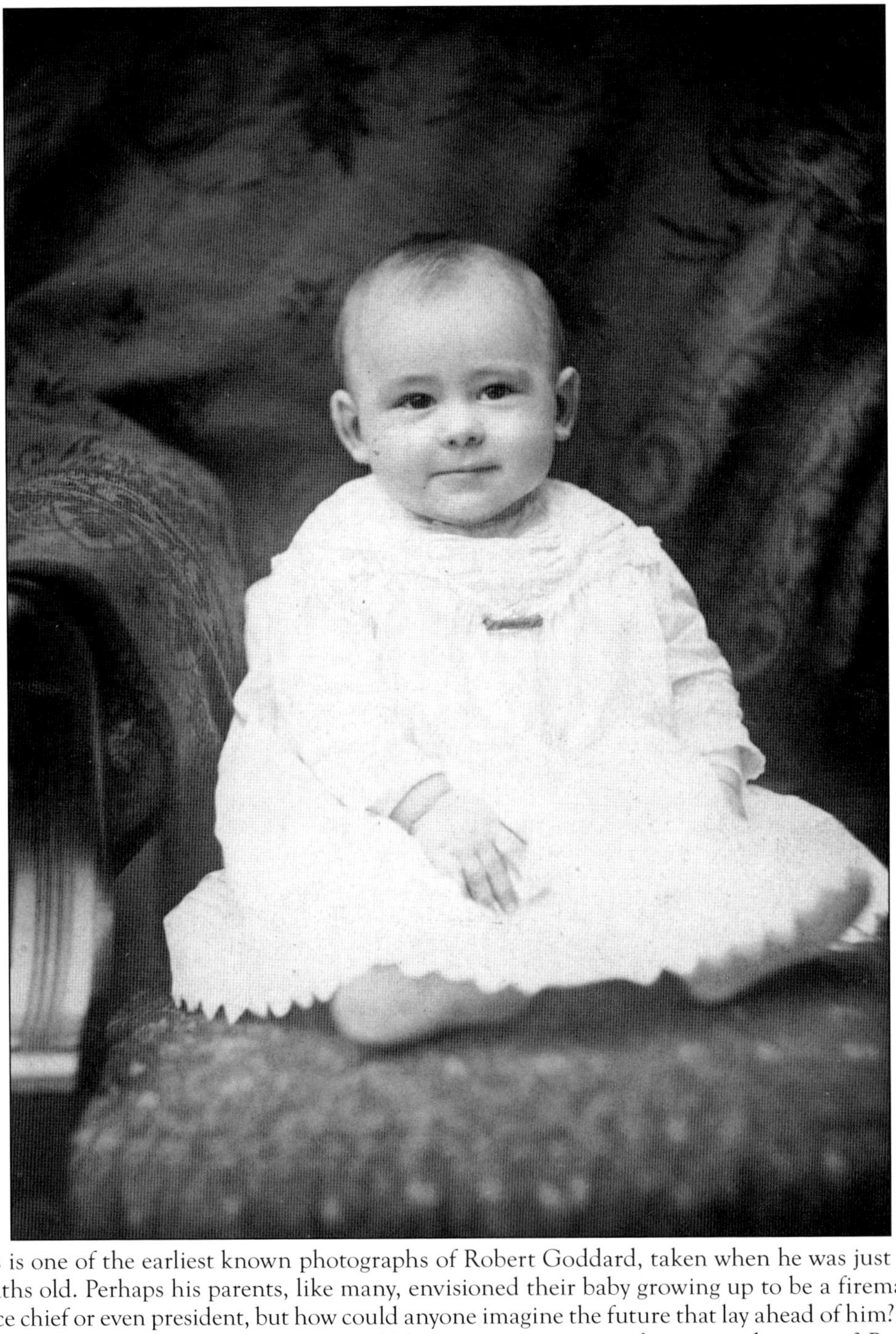

This is one of the earliest known photographs of Robert Goddard, taken when he was just eight months old. Perhaps his parents, like many, envisioned their baby growing up to be a fireman or police chief or even president, but how could anyone imagine the future that lay ahead of him? How could anyone have suspected that he would become a pioneering physicist and inventor? Perhaps his parents envisioned him being a father one day, but certainly not the father of a revolutionary innovation like space travel. In this image, he is simply a baby, years away from the moment under a cherry tree that would inspire his lifelong pursuit of spaceflight, a legacy that inspired later generations of space pioneers and others around the world. This image marks the beginning of a life that forever changed how humanity reaches for the stars.

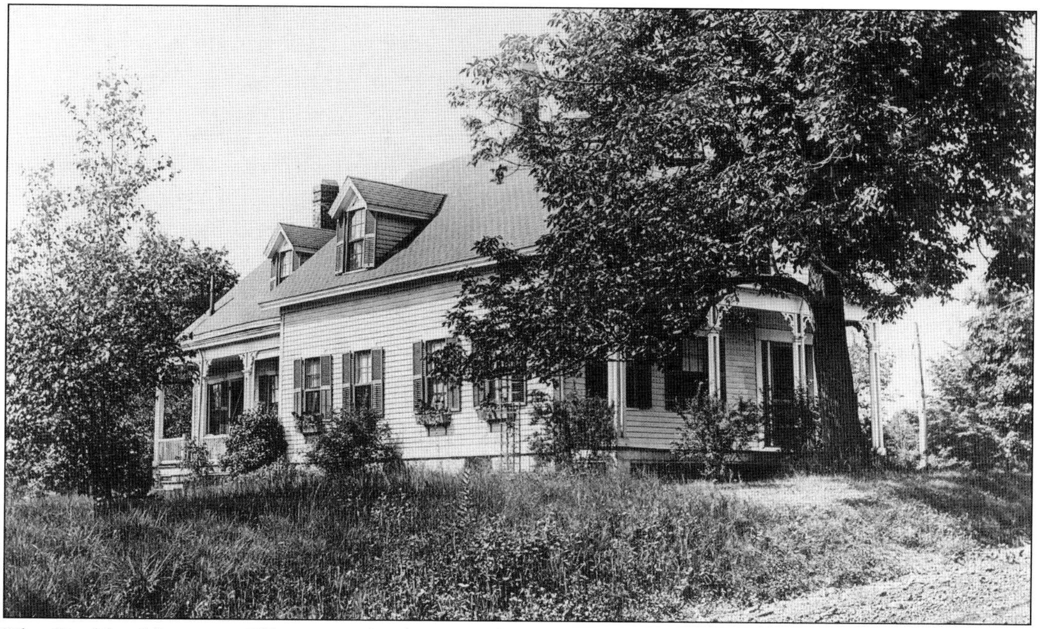

The Goddard home in Worcester, built around 1825, was owned by the family for more than 150 years. It was here that Robert was born in a second-floor bedroom in 1882, where he experienced his cherry tree dream in 1899, where he lived when he launched the world's first liquid-fueled rocket in 1926, and where Esther lived the last decades of her life after Robert passed. They remodeled and modernized the house in the 1920s, and other changes followed in later years, such as the enclosure of the patio. The next several pairs of pictures compare views of the house taken several decades apart and chronicle some of the changes evident from the outside, including replacement of chimneys, narrowing of the front porch, and expansion of the patio. Though the exact dates of many photographs are uncertain, comparisons reveal clear transformations. The view here is from Tallawanda Drive.

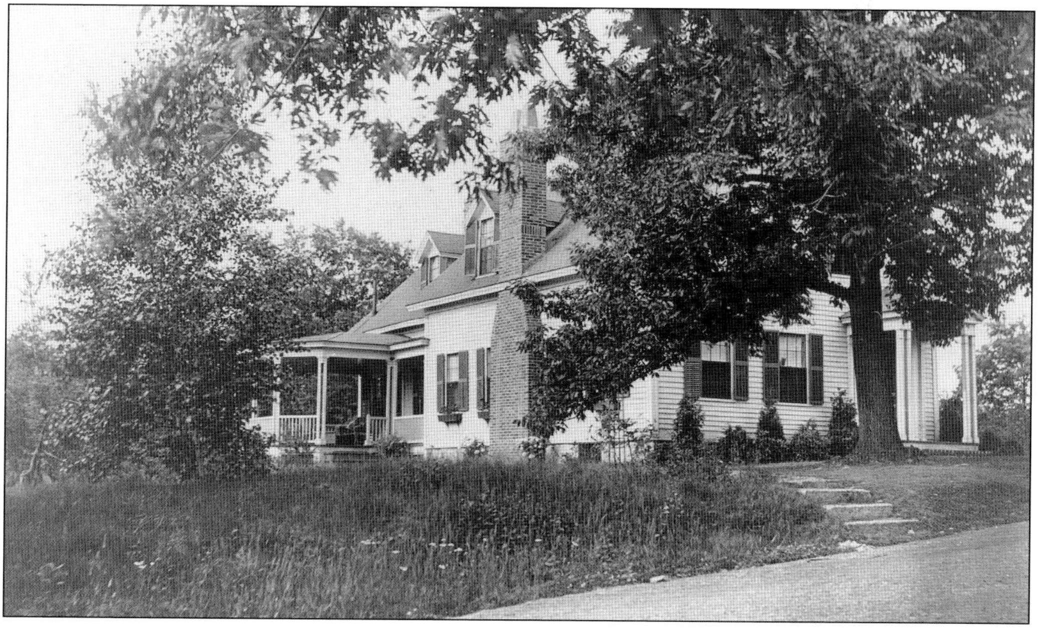

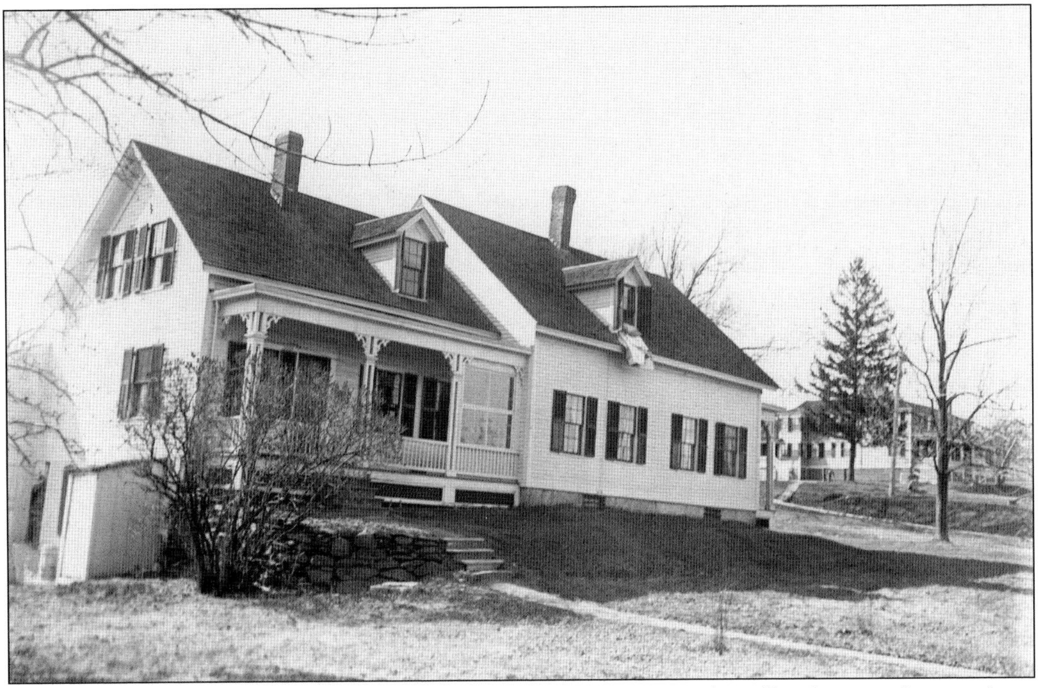

The Goddard home is located on what was once known as Maple Hill. The view here is from the back yard, with Tallawanda Drive coming in from the right. The end of the front porch, still retaining its original width, is just visible in the first picture; it had been narrowed by the time the second picture was taken. The change in chimneys is striking, from a smaller one at the peak of each segment of the house to a single large one that shows through the edge of the roof. The patio, partly obscured by vegetation, is also expanded.

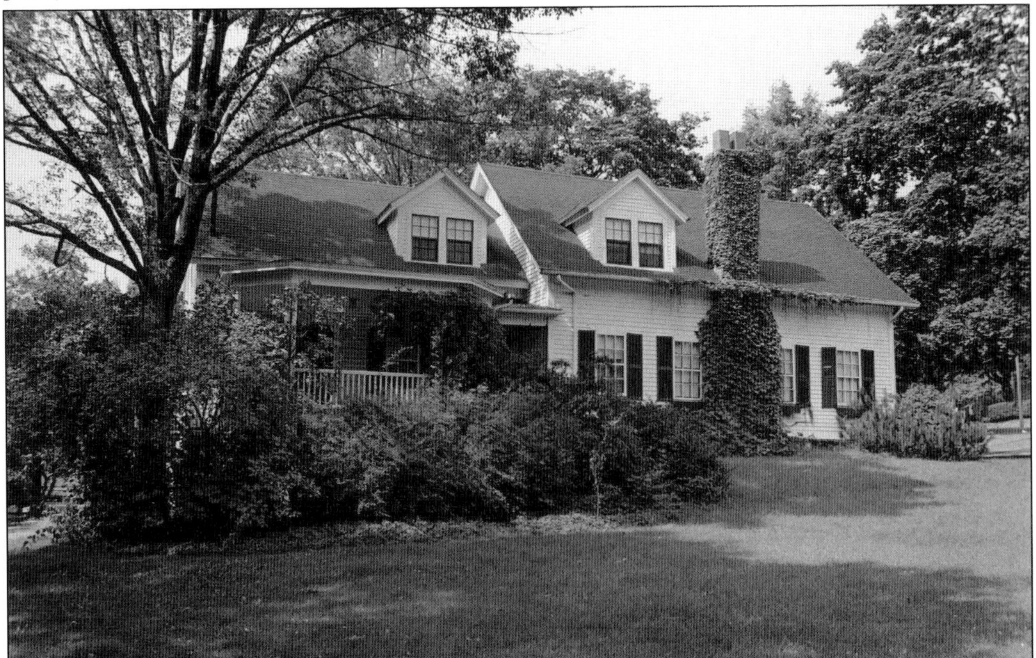

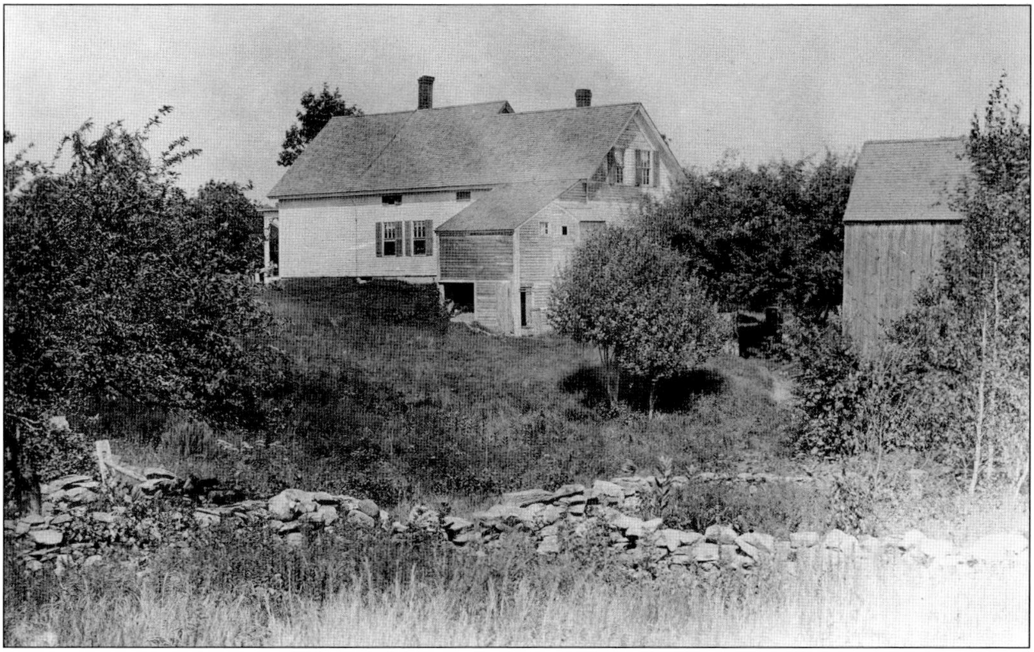

Robert Goddard's cherished cherry tree, the one from which he experienced his transformative "Anniversary Day" vision in 1899, was destroyed by the Great New England Hurricane of 1938. Historians have noted that he entered the following statement in his diary (though examination of the entry shows it to be illegible): "Cherry tree destroyed by hurricane. Must go on alone." The exact location of the tree remains uncertain, but Goddard described it in his *Material for an Autobiography* as "at the back of the barn." Pictures of the tree, seen later in this book, indicate it was also near a brook on the downside of a slope. The first image here shows the barn as it once stood in relation to the Goddard house, viewed from near modern-day Brookline Street. The second image offers a different vantage point, likely from the backyard, with the brook in the foreground (note the house to the left is not the Goddard residence).

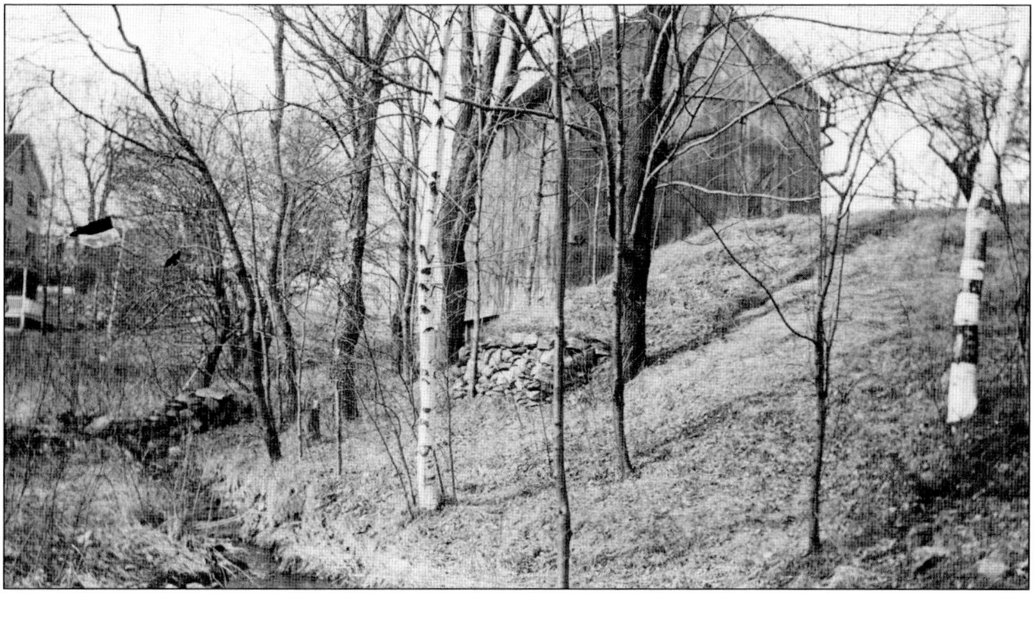

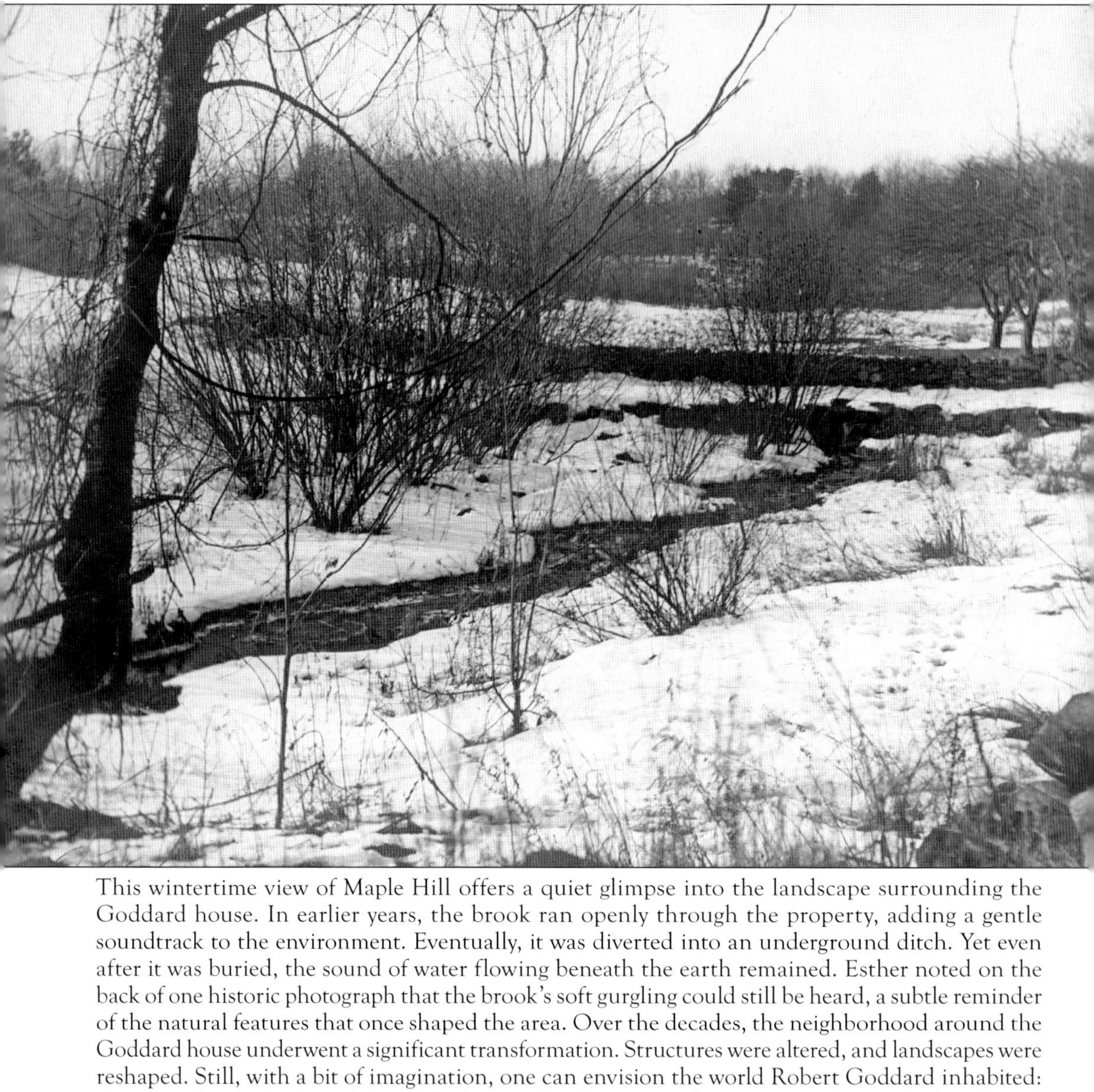

This wintertime view of Maple Hill offers a quiet glimpse into the landscape surrounding the Goddard house. In earlier years, the brook ran openly through the property, adding a gentle soundtrack to the environment. Eventually, it was diverted into an underground ditch. Yet even after it was buried, the sound of water flowing beneath the earth remained. Esther noted on the back of one historic photograph that the brook's soft gurgling could still be heard, a subtle reminder of the natural features that once shaped the area. Over the decades, the neighborhood around the Goddard house underwent a significant transformation. Structures were altered, and landscapes were reshaped. Still, with a bit of imagination, one can envision the world Robert Goddard inhabited: the quiet slopes, the barn behind the house, the cherry tree that once stood near the brook, and a night sky full of stars and unpolluted by city lights.

In about 1890, several generations of Goddards gathered at the front porch of the Goddard house at 1 Tallawanda Drive on Maple Hill. From left to right are Robert's mother, Fannie, paternal grandmother, Mary Pease Upham Goddard, Robert, Robert's great-grandmother Elvira Goddard Ward, and his father, Nahum. Through the years, these and other members of the family lived in the two-story house.

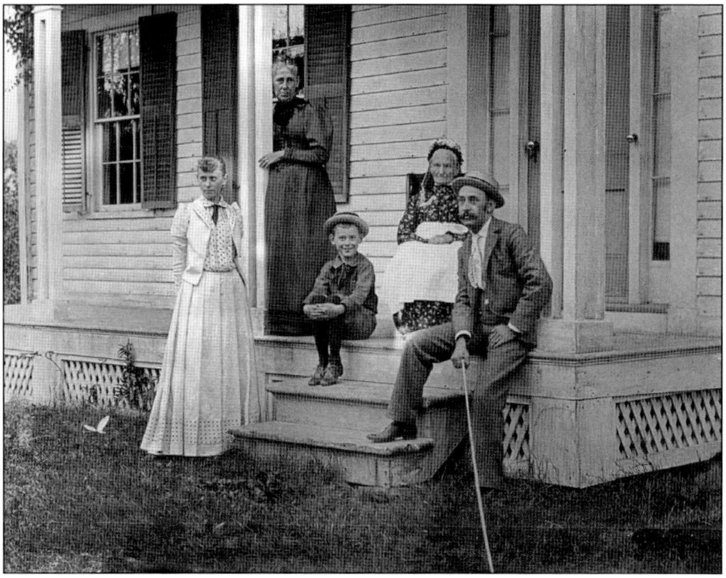

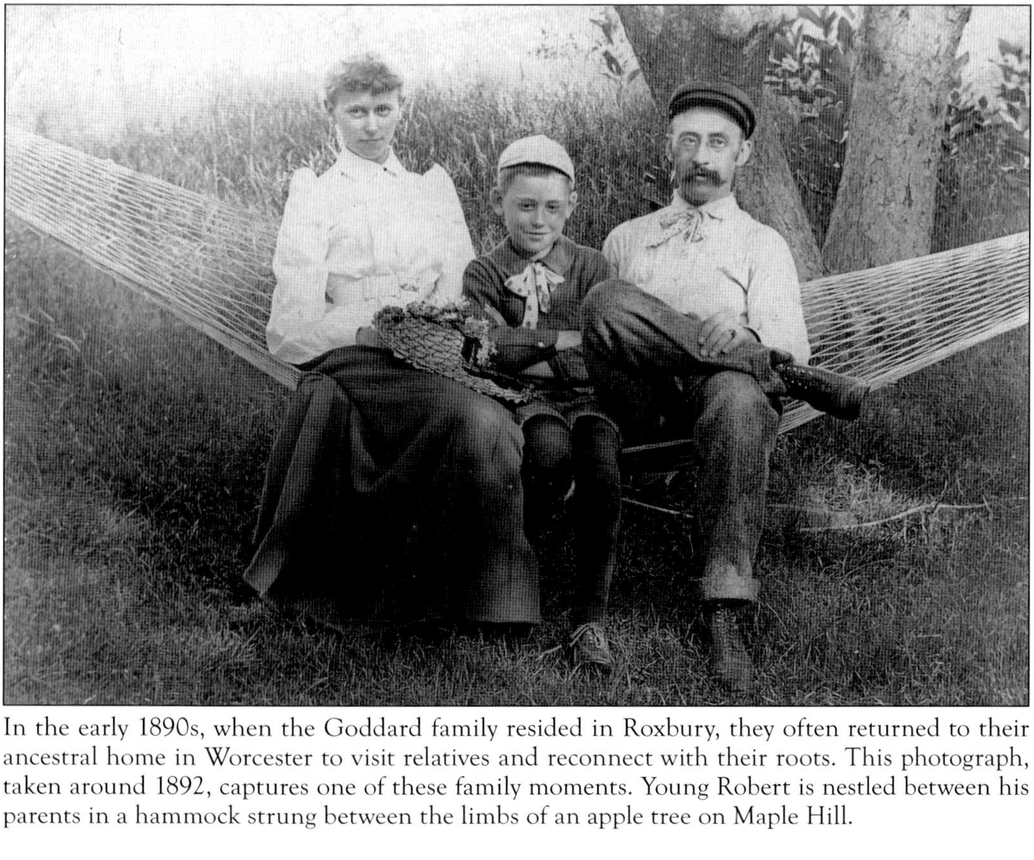

In the early 1890s, when the Goddard family resided in Roxbury, they often returned to their ancestral home in Worcester to visit relatives and reconnect with their roots. This photograph, taken around 1892, captures one of these family moments. Young Robert is nestled between his parents in a hammock strung between the limbs of an apple tree on Maple Hill.

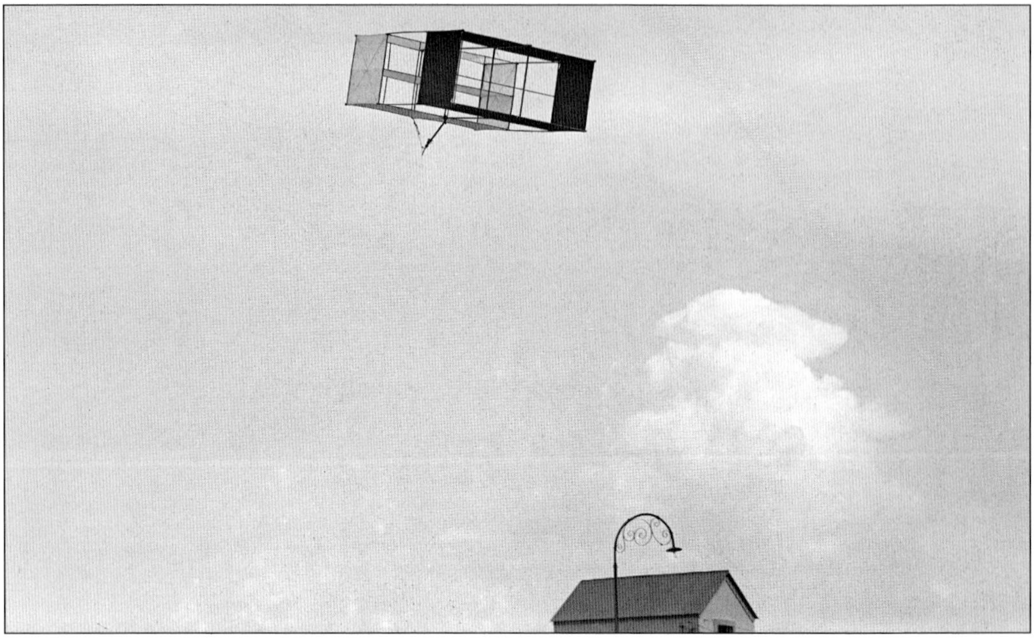

Like most generations of youth, Goddard was a child of his times and enjoyed delving into fads or otherwise fashionable activities. He took up archery for a time, and as bicycling surged in popularity during the 1890s, he eagerly joined the trend. Another fascination was with box kites, invented by Australian Lawrence Hargrave in 1893. Goddard went through a phase of building and flying these, building on his longtime fascination with the sky. (Library of Congress.)

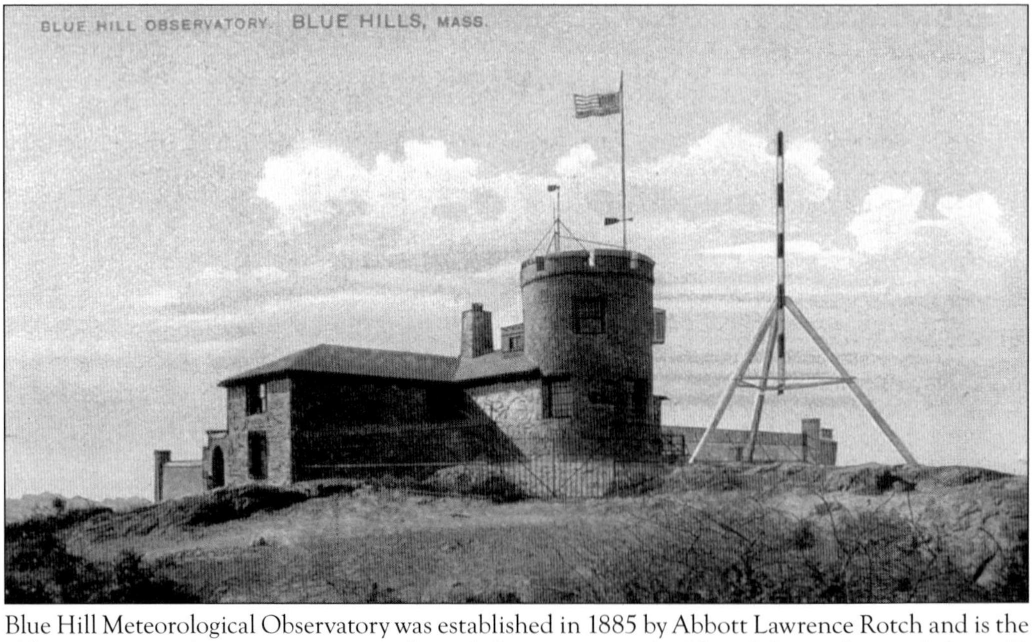

Blue Hill Meteorological Observatory was established in 1885 by Abbott Lawrence Rotch and is the oldest continuously operating meteorological observatory in the United States. Located about 50 miles east of Worcester and 17 miles south of Roxbury, it was close enough for the Goddard family to visit. It may well have helped amplify Goddard's captivation with the skies. (Kevin Schindler personal collection.)

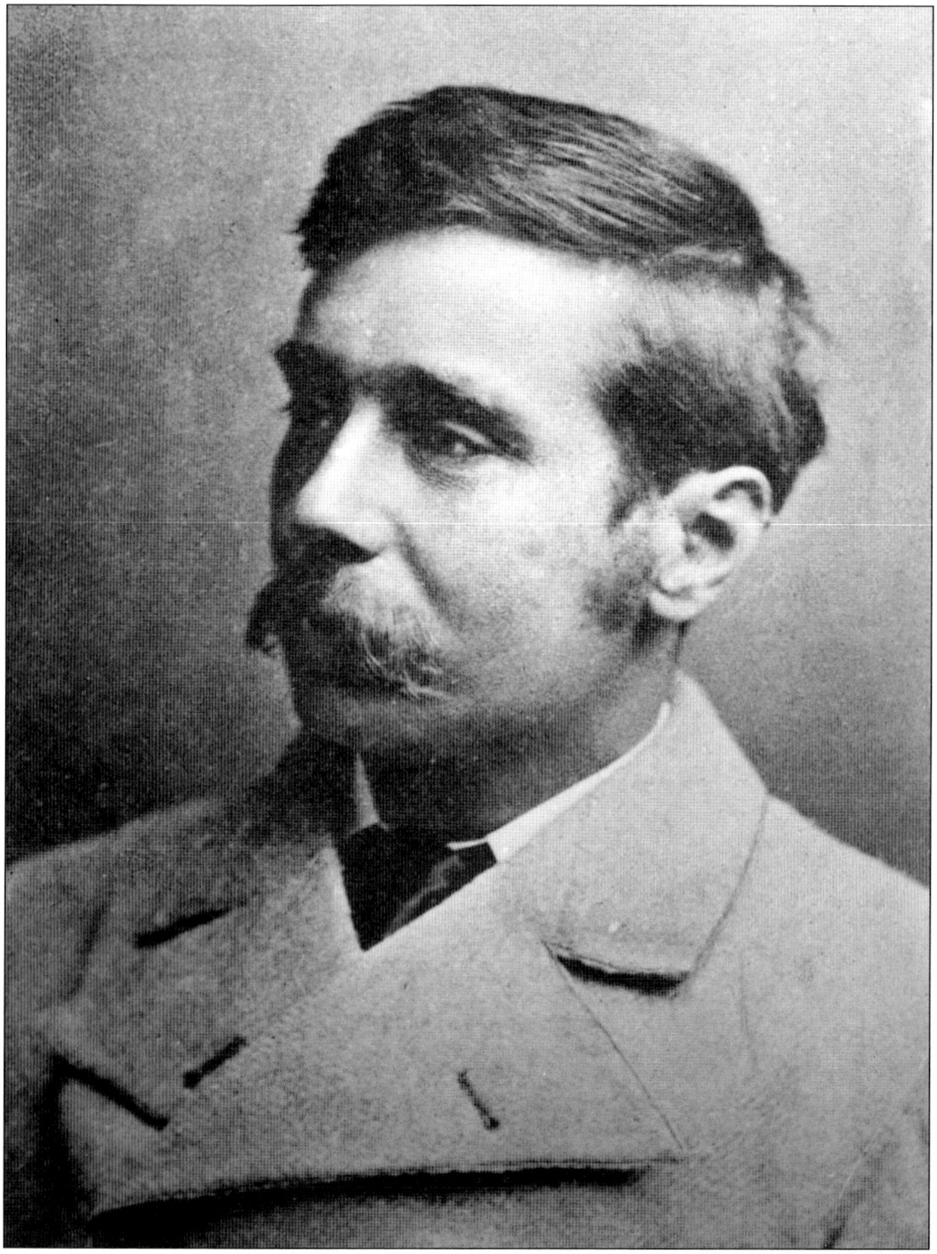

From a young age, Robert Goddard developed a deep love of reading and explored a wide variety of genres, including both nonfiction and fiction. He was often drawn to writers who blended imagination with science, and two authors stood out as lifelong inspirations: Jules Verne and H.G. Wells. Verne, the French novelist famous for incorporating the latest scientific knowledge into his stories, wrote classics like *From the Earth to the Moon*, which fascinated Goddard and sparked his early interest in space travel. Although Verne published most of his iconic works before Goddard was born, their influence remained strong throughout his youth. Wells, shown here, published his books during Goddard's formative years and also had a profound impact on his career. One book in particular, *The War of the Worlds*, caught Goddard's attention and helped inspire the course of his professional life. (Library of Congress.)

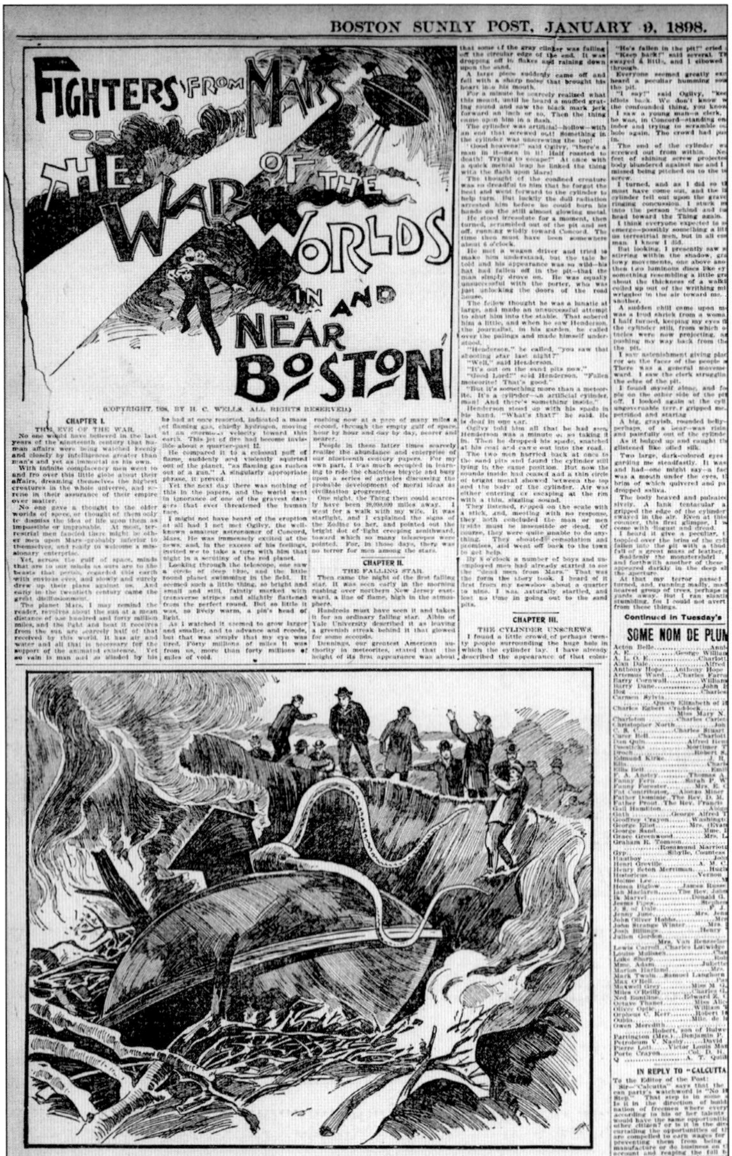
BOSTON SUNDAY POST, JANUARY 9, 1898.

FIGHTERS FROM MARS OR THE WAR OF THE WORLDS IN AND NEAR BOSTON

(COPYRIGHT, 1898, BY H. G. WELLS. ALL RIGHTS RESERVED.)

CHAPTER I.
THE EVE OF THE WAR.

CHAPTER II.
THE FALLING STAR.

CHAPTER III.
THE CYLINDER UNSCREWS.

Continued in Tuesday's

SOME NOM DE PLUMES

IN REPLY TO "CALCUTTA"

THE MONSTER EMERGING FROM THE WAR CYLINDER AT CONCORD, MASS.

Beginning on January 9, 1898, the *Boston Post* ran a serialized adaptation of H.G. Wells's recently published science fiction book *The War of the Worlds*, retitled *Fighters from Mars, or the War of the Worlds in and near Boston*. This unauthorized version relocated the Martian invasion from southern England to the Boston area, altering key scientific details and simplifying the narrative for local readers. For a young Robert Goddard, this localized version of Wells's story was transformative. The vivid imagery of Martians descending on familiar streets captured his imagination and sparked a lifelong fascination with space travel. Alongside its sequel, *Edison's Conquest of Mars*, the story played a pivotal role in inspiring Goddard's early interest in rocketry and interplanetary exploration. The following year, this Martian imagery was in his mind when, on a cool October evening, he experienced his cherry tree dream that inspired his vision of developing the means of traveling into space. (Kevin Schindler personal collection.)

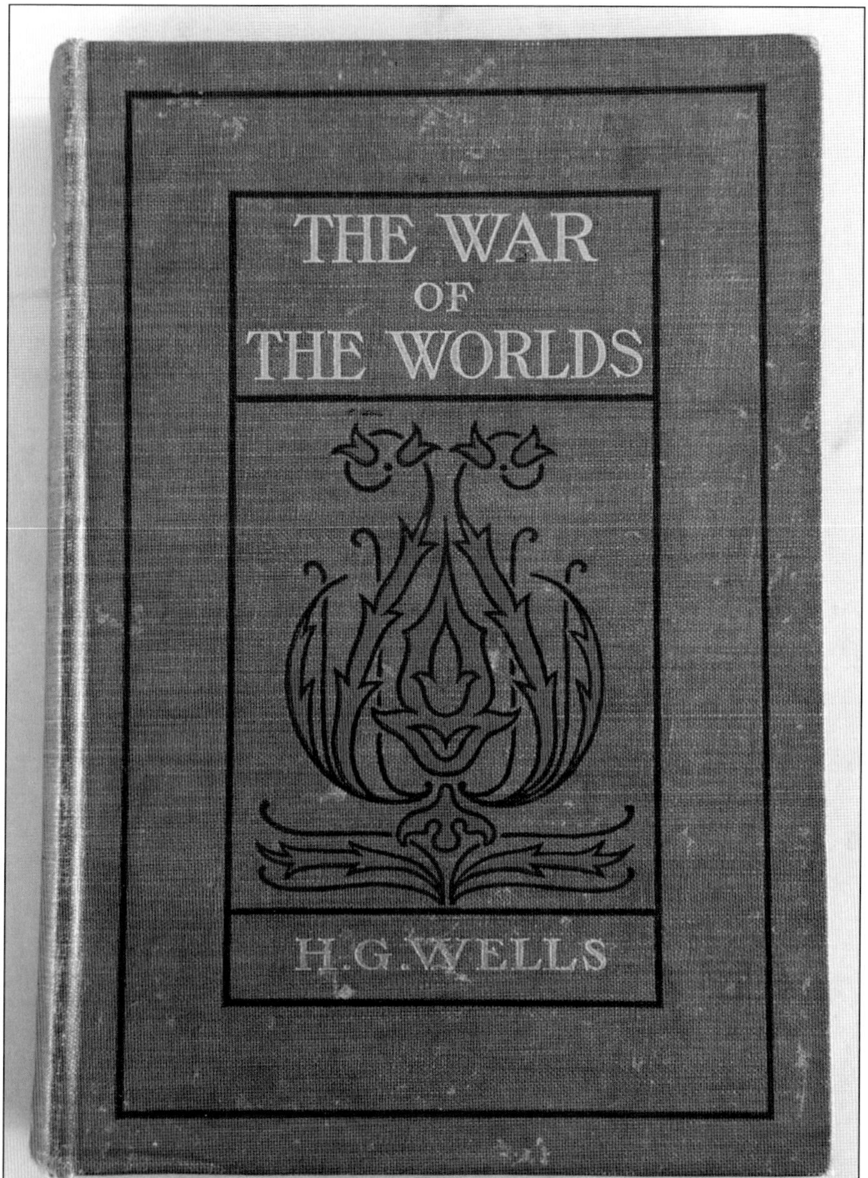

After his initial reading of the Boston-based serialized version of *The War of the Worlds*, Goddard read it in book form, over and over, through the years. H.G. Wells's vivid depiction of a Martian invasion and advanced alien technology stirred Goddard's imagination and helped crystallize his vision of space travel. The story became a personal touchstone, reminding him why he pursued rocketry despite skepticism and ridicule from the press and the scientific community. Goddard was not alone in finding inspiration in Wells's work. On Halloween night in 1938, Orson Welles famously adapted *The War of the Worlds* for radio. He presented it as a series of realistic news bulletins that resulted in widespread panic among listeners who believed the broadcast represented a real event. The story has since been adapted into multiple motion pictures and remains a cornerstone of science fiction. For Goddard, it was more than science fiction, it was fuel for a lifelong pursuit of the stars. (Kevin Schindler personal collection.)

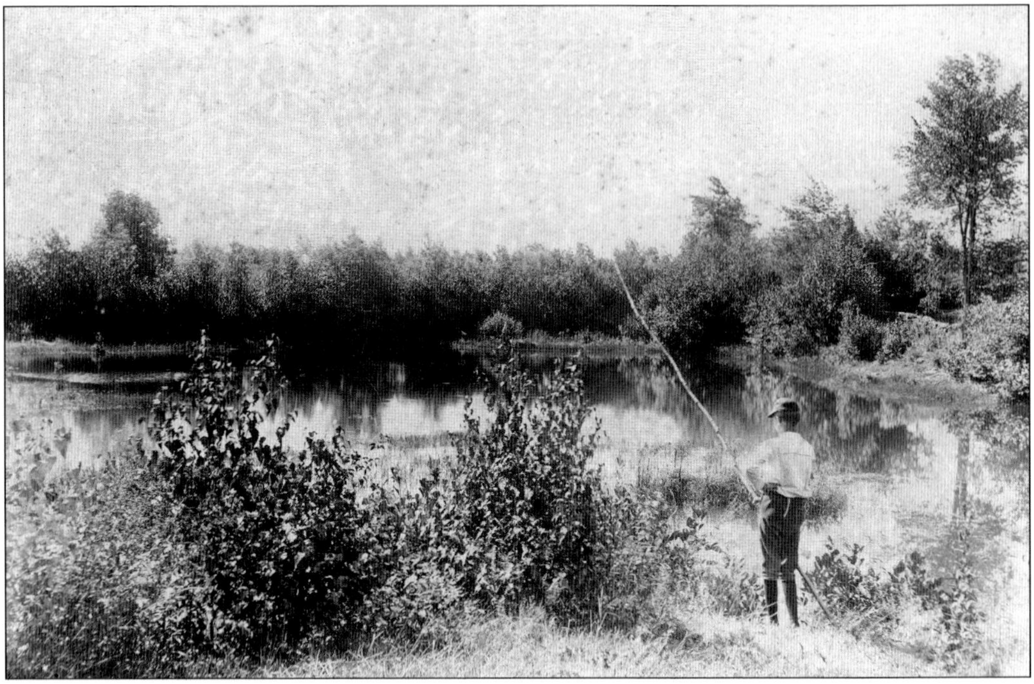

Located less than two miles northwest of the Goddard house, along a road known in modern times as Goddard Memorial Drive, was Parson's Pond, a man-made pond designed to power a cider mill. Young Goddard enjoyed fishing here—as seen in this image—as well as at other local fishing holes.

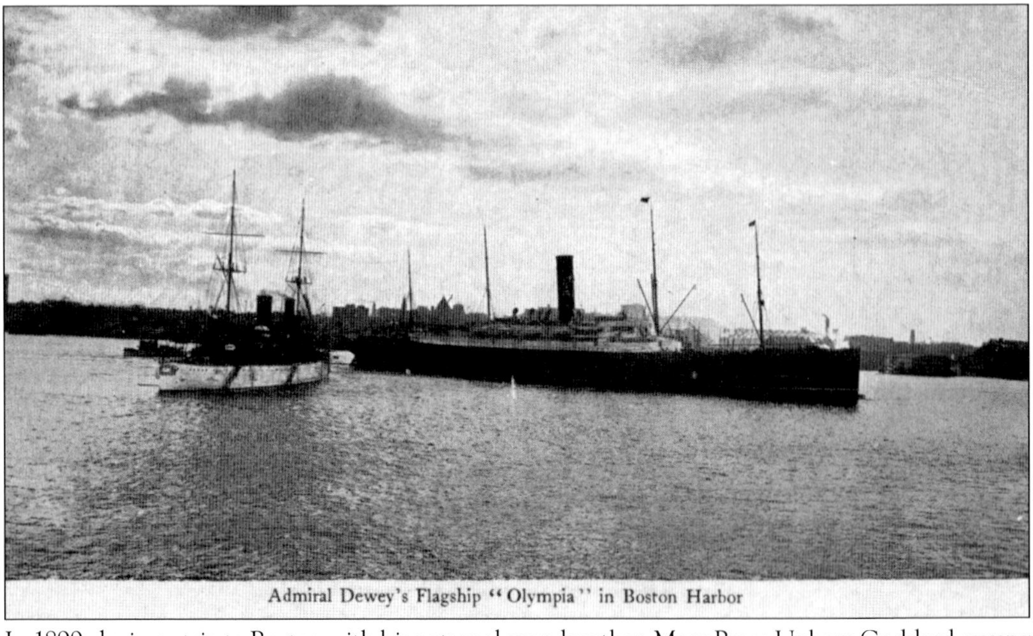

In 1899, during a trip to Boston with his paternal grandmother, Mary Pease Upham Goddard, young Robert Goddard experienced a rare and unforgettable moment: dining aboard Adm. George Dewey's flagship, the USS *Olympia*. The ship had recently returned from its celebrated role in the Spanish-American War. The opportunity to dine aboard such a historic vessel came courtesy of William Long, a family friend and assistant paymaster in the US Navy. (Kevin Schindler personal collection.)

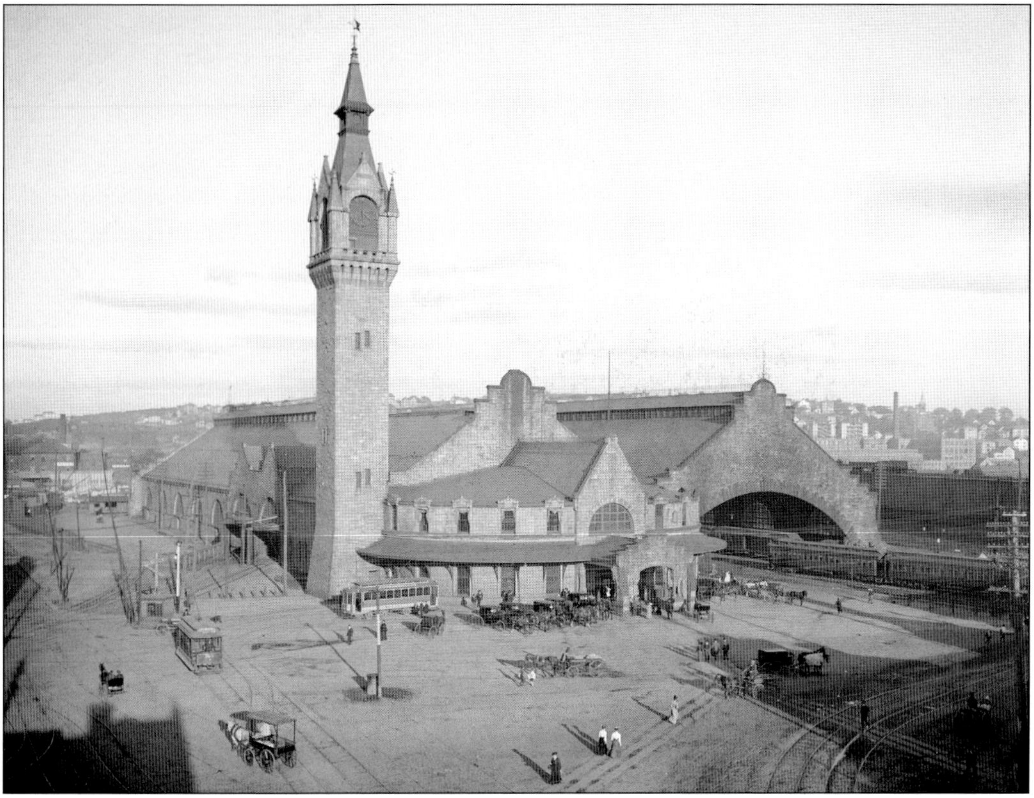

For trips to Boston and beyond, Goddard could board a train at Worcester's original Union Station, built in 1875 (above). The station featured a 21-story-tall clock tower that was visible from miles in every direction. During the latter days of Goddard's college years, in 1911, the station was replaced with a twin-towered structure of marble that rose 175 feet in the air. In 1926, the year of Goddard's historic first launch of a liquid-fueled rocket, the towers were razed due to instability caused by the rattling of passing trains. The station itself remained in operation, and Goddard continued using it to travel back and forth from Worcester for the rest of his life. After years of falling into disrepair, the station was abandoned in the 1970s, but it was restored a quarter century later, with new towers built with structurally sound material and new engineering techniques. (Above, Museum of Worcester; below, Kevin Schindler personal collection.)

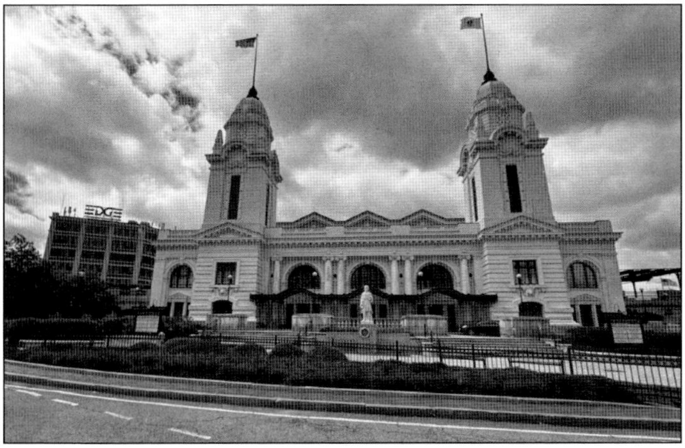

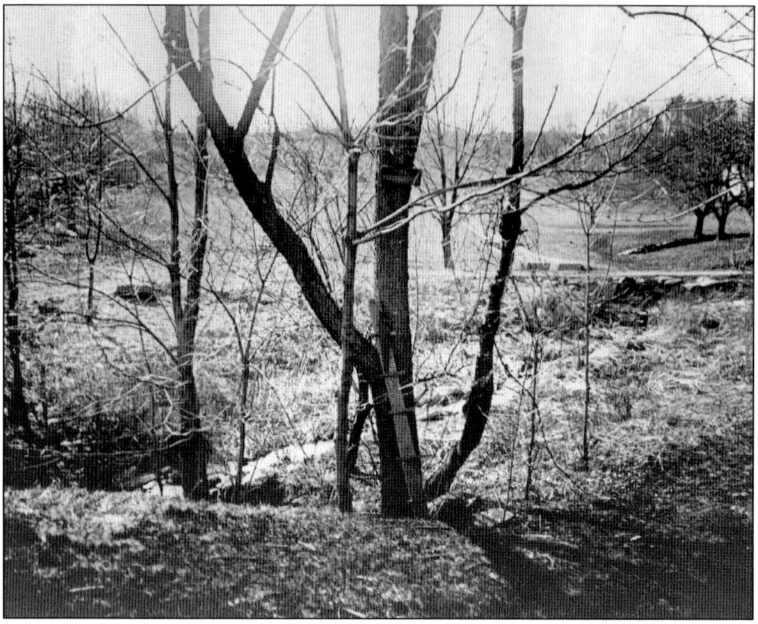

On October 19, 1899, Goddard experienced the celebrated epiphany that changed the course of his life. While climbing a cherry tree behind the barn to trim branches, he had a dreamlike vision of devising a method of travelling to space, specifically to Mars. This picture, which he took years later, shows a ladder, with a saw balanced on top, leaning against the tree, along with a no trespassing sign above them.

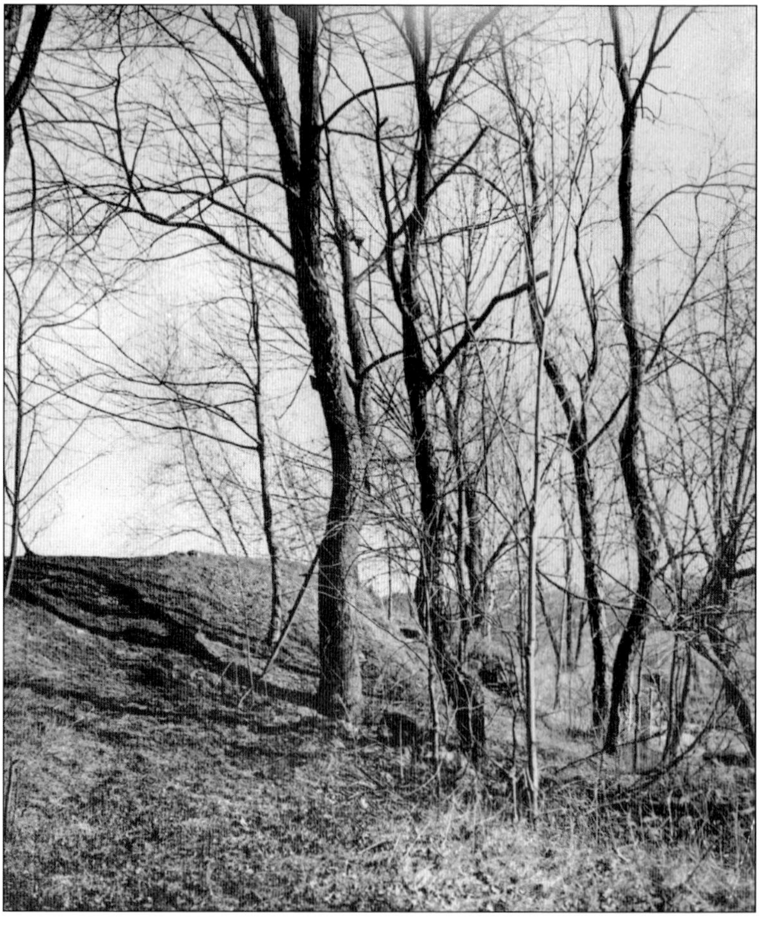

Goddard would forever refer to October 19 as "Anniversary Day" and celebrate it each year. As he recalled in his *Material for an Autobiography*, "I was a different boy when I descended the ladder. Life now had a purpose for me." The experience inspired him to study rocketry in order to turn his dream of space travel into reality. Here is another angle of the tree with the leaning ladder.

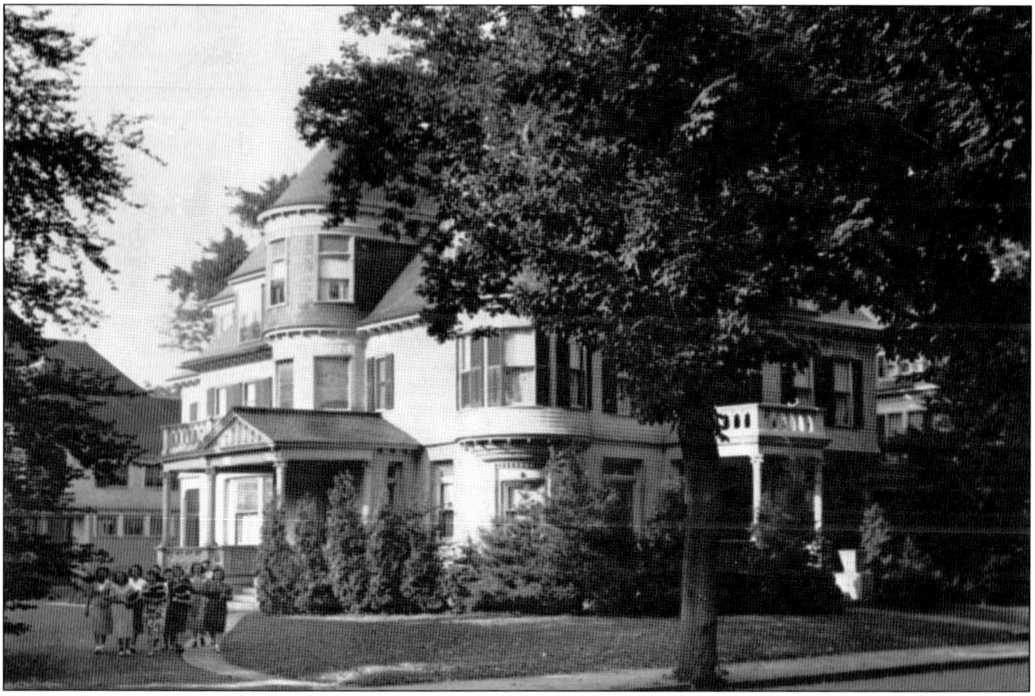

After moving back to Worcester from Roxbury in 1898, Goddard was unable to return to high school due to lingering health issues. In the fall of 1900, feeling better but still not well enough for full-time schooling, he enrolled at the Worcester campus of Becker's Business College, a commercial arts institution. However, this brief attempt to resume formal education was cut short when his health deteriorated again, forcing him to withdraw. (Museum of Worcester.)

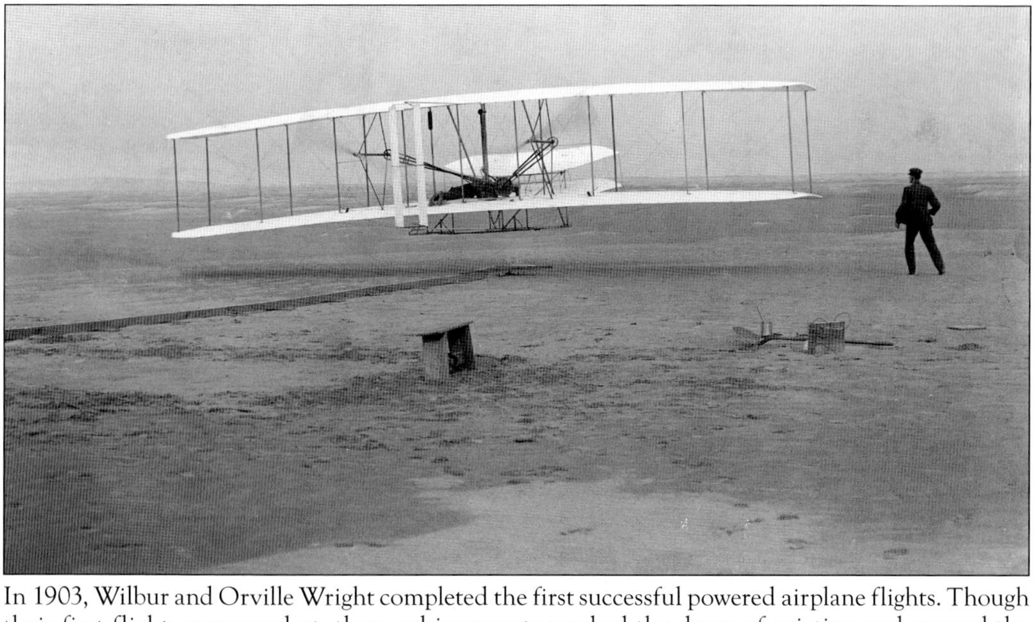

In 1903, Wilbur and Orville Wright completed the first successful powered airplane flights. Though their first flights were modest, these achievements marked the dawn of aviation and opened the door to the future of air travel. Twenty-three years later, Goddard extended humanity's vision of flight beyond Earth's atmosphere with his historic rocket launch. (Library of Congress.)

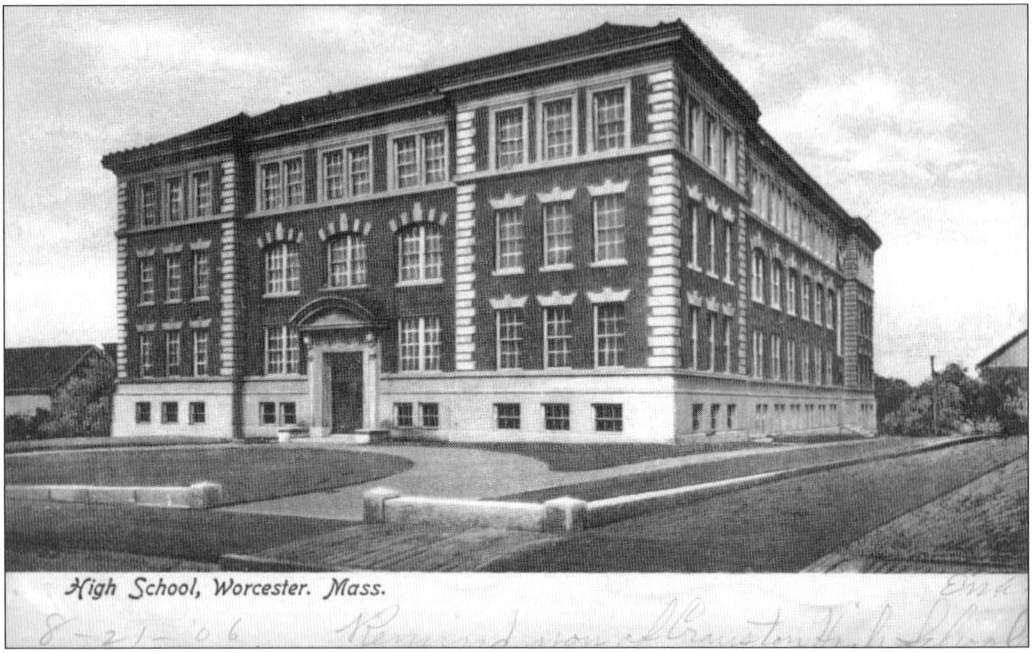

When the Wright brothers took to the skies, Goddard was back in school. Finally healthy, he had enrolled as a sophomore at South High in September 1901, just a month before his 19th birthday. While his illnesses were not permanently behind him, they never again caused him to miss significant time in high school or later in college. (Kevin Schindler personal collection.)

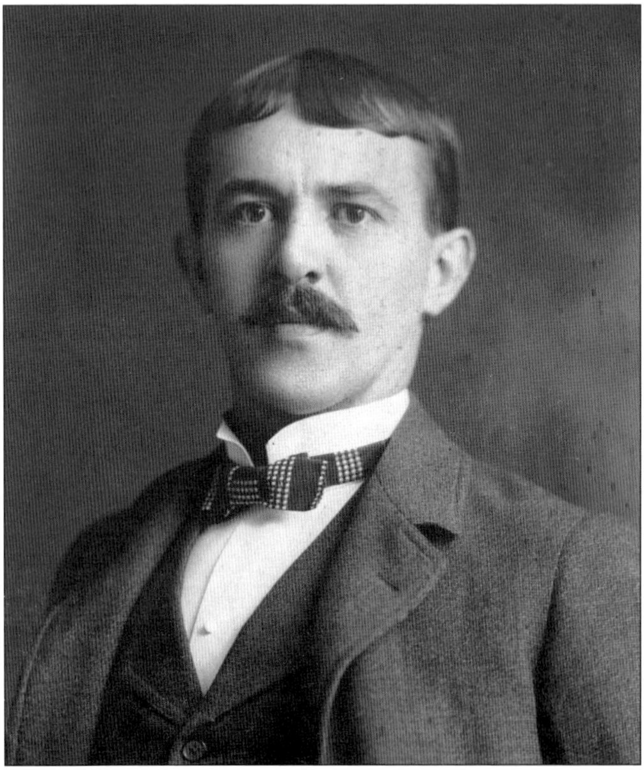

Two of the most influential educators in Robert Goddard's early life were his math teacher Miss Fields and his physics teacher Calvin Andrews, both of South High. Andrews, shown here, introduced the inquisitive Goddard to the principles of mechanics and dynamics, nurturing his fascination with the physical world. More than just a teacher, Andrews became a trusted mentor and sounding board for Goddard's imaginative ideas, many of which would later shape the future of rocketry.

At South High, Goddard did not confine his energy to the classroom. Alongside his academic pursuits, he served as class president, developed his lifelong interest in drawing, and even composed the graduation song for the class of 1904. True to the visionary mindset that would define his life, Goddard included a line that encouraged his classmates to persevere through adversity: "Take the path that leads us onward e'en when clouds our course obscure."

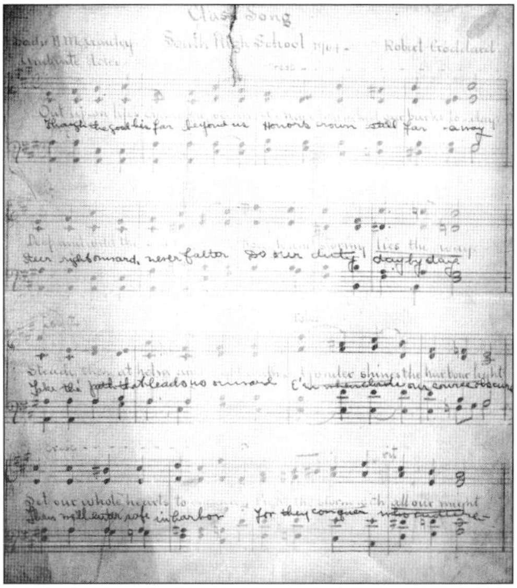

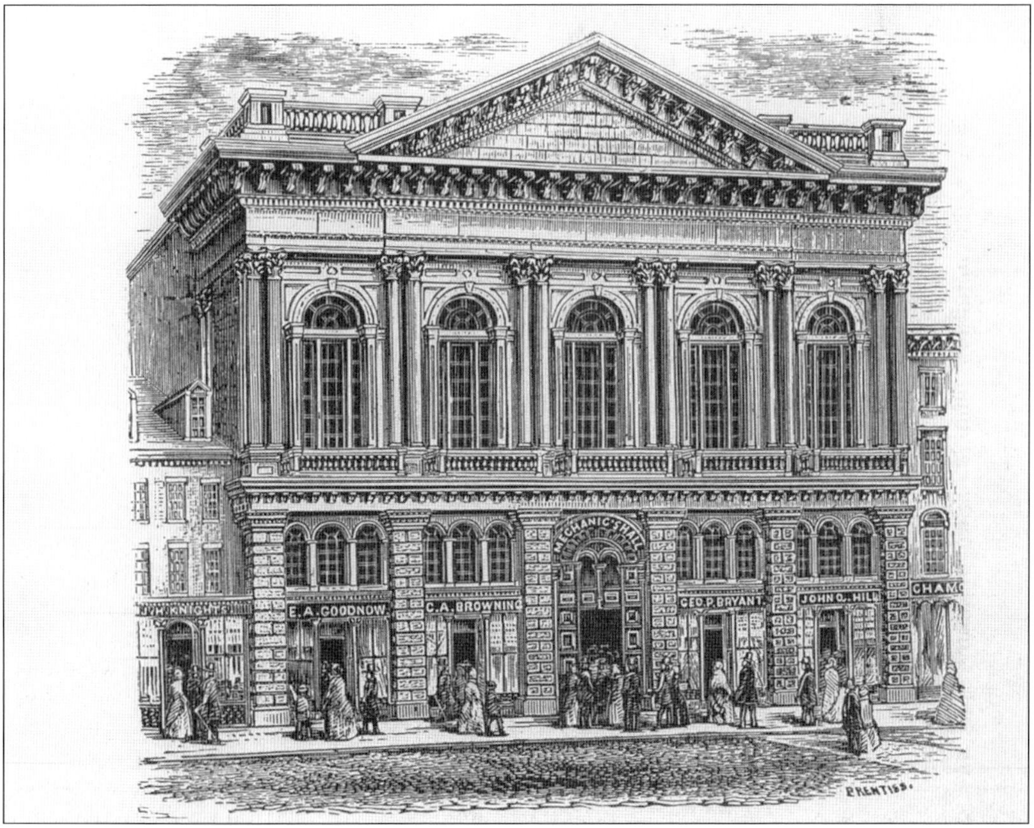

Goddard excelled academically at South High, and when it came time for graduation in 1904, his English teacher asked him to give a speech as the top graduating male. The top graduating female was Miriam Olmstead, and she would also present a speech. Graduation would take place at Mechanics Hall, a majestic concert hall in downtown Worcester that had been funded by community industrial leaders. (Museum of Worcester.)

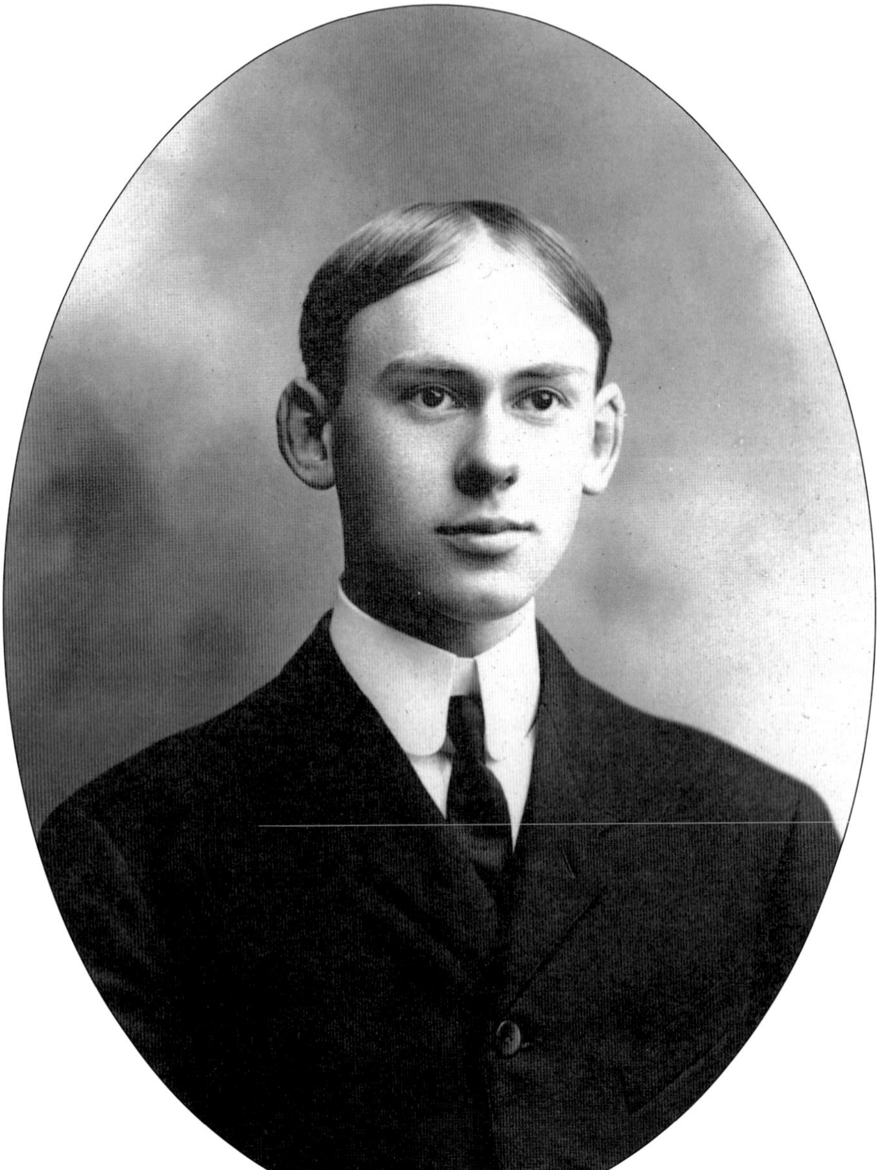

Goddard was 21 years old when he graduated from South High School in 1904, several years older than his classmates due to the extended absences caused by illness earlier in his life. While his high school career was behind him, he had much to look forward to with the prospect of college as well as a girl in his life. She was none other than Miriam Olmstead, whom Goddard had gotten to know as the two prepared for their graduation ceremony. Their shared responsibilities and mutual admiration blossomed into a romantic relationship and eventually an engagement. However, life soon pulled them in different directions. Goddard enrolled at Worcester Polytechnic Institute, while Miriam attended Smith College in Northampton. The physical distance between them, combined with Goddard's decision not to marry until he had completed his education and saved at least $1,000, created emotional strain. Over time, the relationship quietly unraveled, and they finally went their own ways.

Goddard's high school graduation program indicates that three schools—Classical High, English High, and South High—participated, which may explain why the ceremony was held at the spacious Mechanics Hall. The final paragraph of Goddard's "On Taking Things for Granted" speech, including the celebrated final phrase that scholars and students have repeatedly quoted through the years, follows: "But, just as in the sciences we have learned that we are too ignorant safely to pronounce anything impossible, so for the individual, since we cannot know just what are his limitations, we can hardly say with certainty that anything is necessarily within or beyond his grasp. Each must remember that no one can predict to what heights of wealth, fame, or usefulness he may rise until he has honestly endeavored, and he should derive courage from the fact that all sciences have been, at some time, in the same condition as he, and that it has often proved true that the dream of yesterday is the hope of today and the reality of tomorrow."

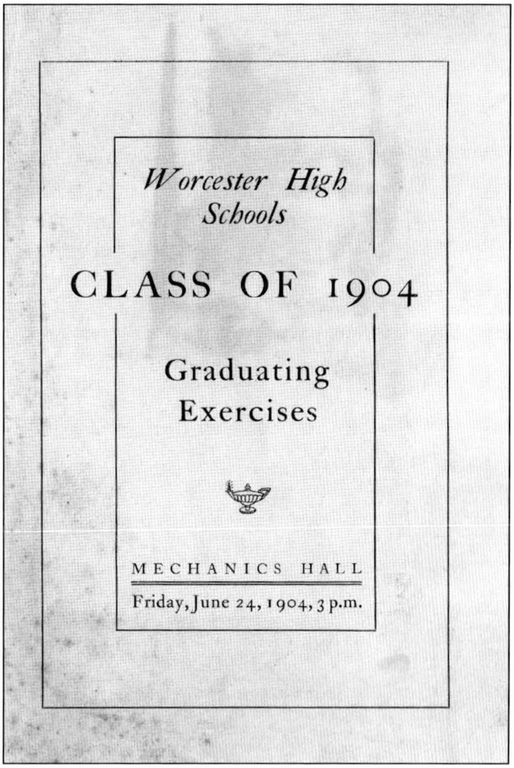

Worcester High Schools

CLASS OF 1904

Graduating Exercises

MECHANICS HALL

Friday, June 24, 1904, 3 p.m.

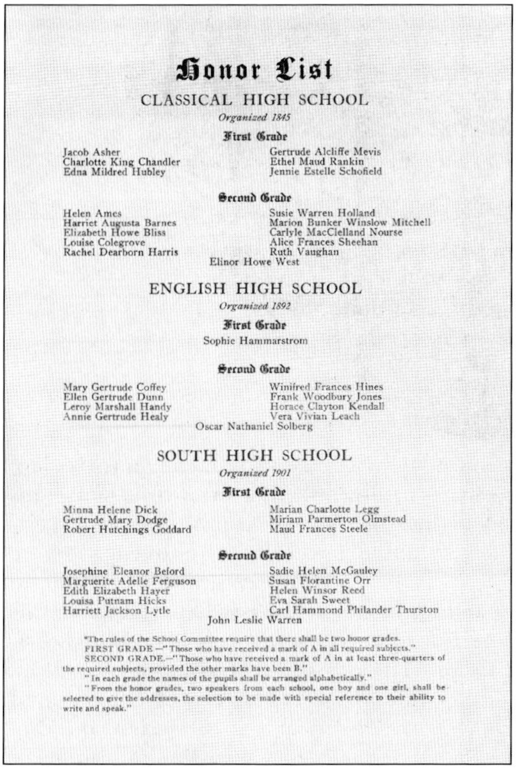

Honor List

CLASSICAL HIGH SCHOOL

Organized 1845

First Grade

Jacob Asher
Charlotte King Chandler
Edna Mildred Hubley
Gertrude Alcliffe Mevis
Ethel Maud Rankin
Jennie Estelle Schofield

Second Grade

Helen Ames
Harriet Augusta Barnes
Elizabeth Howe Bliss
Louise Colegrove
Rachel Dearborn Harris
Susie Warren Holland
Marion Bunker Winslow Mitchell
Carlyle MacClelland Nourse
Alice Frances Sheehan
Ruth Vaughan
Elinor Howe West

ENGLISH HIGH SCHOOL

Organized 1892

First Grade

Sophie Hammarstrom

Second Grade

Mary Gertrude Coffey
Ellen Gertrude Dunn
Leroy Marshall Handy
Annie Gertrude Healy
Winifred Frances Hines
Frank Woodbury Jones
Horace Clayton Kendall
Vera Vivian Leach
Oscar Nathaniel Solberg

SOUTH HIGH SCHOOL

Organized 1901

First Grade

Minna Helene Dick
Gertrude Mary Dodge
Robert Hutchings Goddard
Marian Charlotte Legg
Miriam Parmerton Olmstead
Maud Frances Steele

Second Grade

Josephine Eleanor Beford
Marguerite Adelle Ferguson
Edith Elizabeth Hayer
Louisa Putnam Hicks
Harriett Jackson Lytle
Sadie Helen McGauley
Susan Florantine Orr
Helen Winsor Reed
Eva Sarah Sweet
Carl Hammond Philander Thurston
John Leslie Warren

*The rules of the School Committee require that there shall be two honor grades.
FIRST GRADE.—"Those who have received a mark of A in all required subjects."
SECOND GRADE.—"Those who have received a mark of A in at least three-quarters of the required subjects, provided the other marks have been B."
"In each grade the names of the pupils shall be arranged alphabetically."
"From the honor grades, two speakers from each school, one boy and one girl, shall be selected to give the addresses, the selection to be made with special reference to their ability to write and speak."

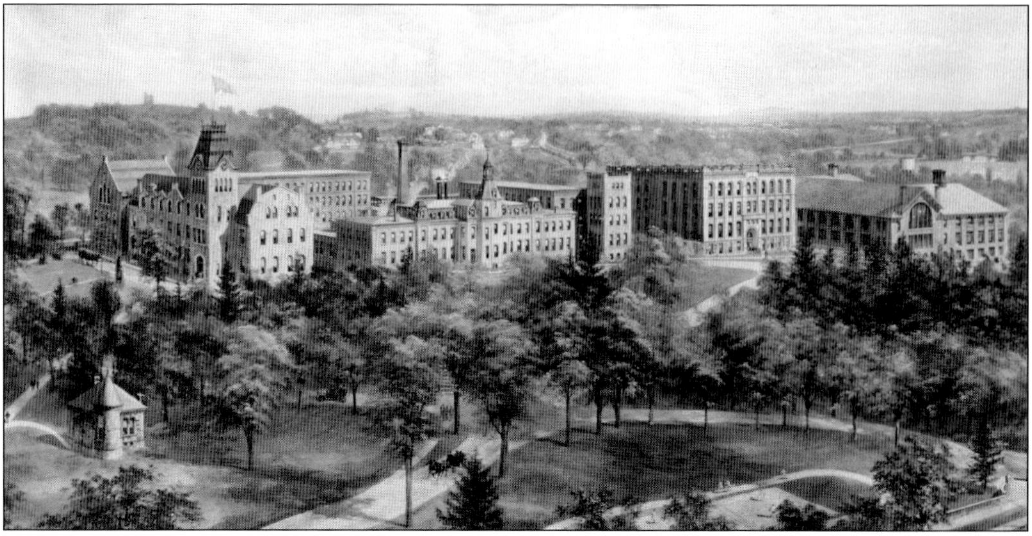

After graduating from South High School in 1904, Goddard enrolled at Worcester Polytechnic Institute (WPI), where he began classes that fall. The campus at the time resembled the 1908 lithograph shown here. Of particular interest is the Magnetic Laboratory, located at the lower left of the image and later known as the "Skull Tomb." Goddard used it for a time after graduation to conduct research.

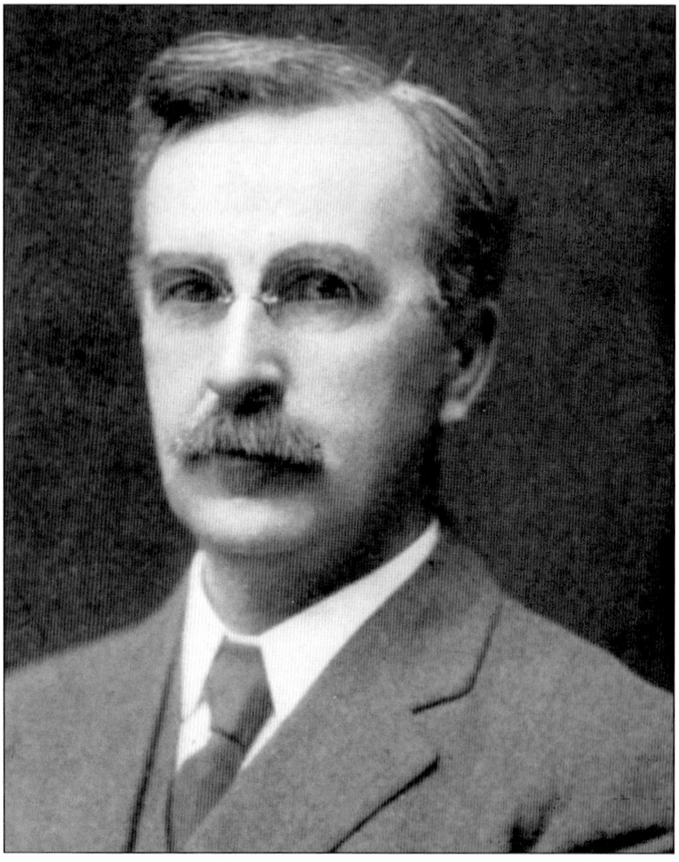

One of Goddard's most valued mentors at WPI was A. Wilmer Duff, an English-born physicist. After stints as a professor in India, England, and Purdue University, he served as professor and head of the physics department at WPI from 1893 to 1936. Goddard served as a lab assistant under Duff and also took on the role of tutor and teacher assistant.

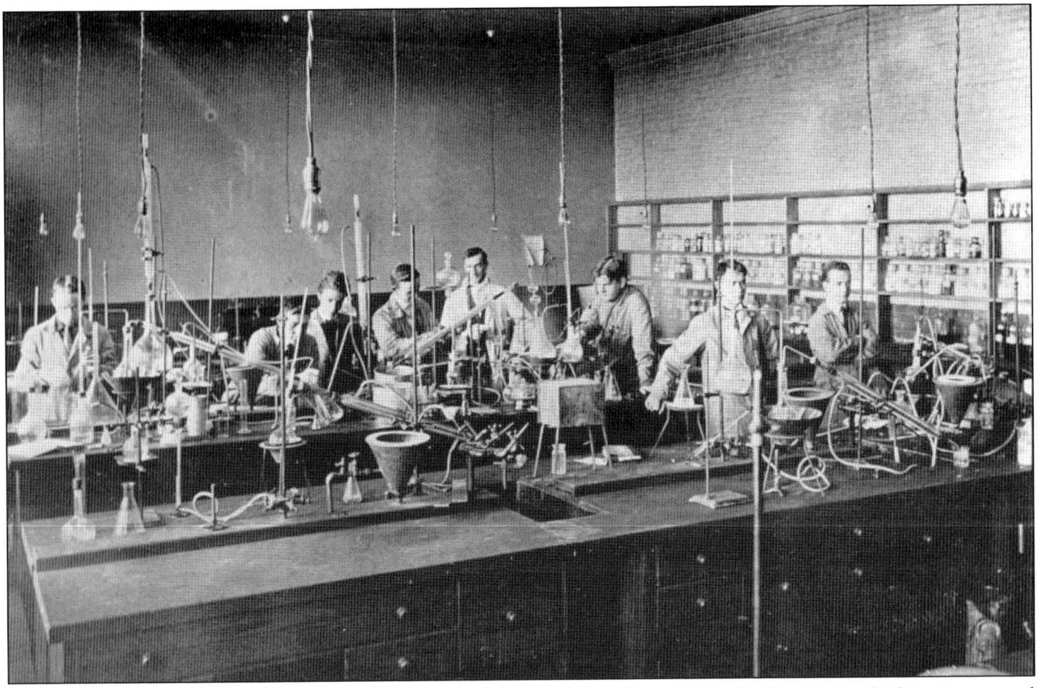

During his time at WPI, Goddard immersed himself in the study of physics and chemistry and spent a lot of time conducting laboratory studies. In the chemistry lab, shown here, Goddard learned important lessons about chemical reactions, combustion, and the properties of gases. He also conducted research in solid fuel propellants before he determined their limitations.

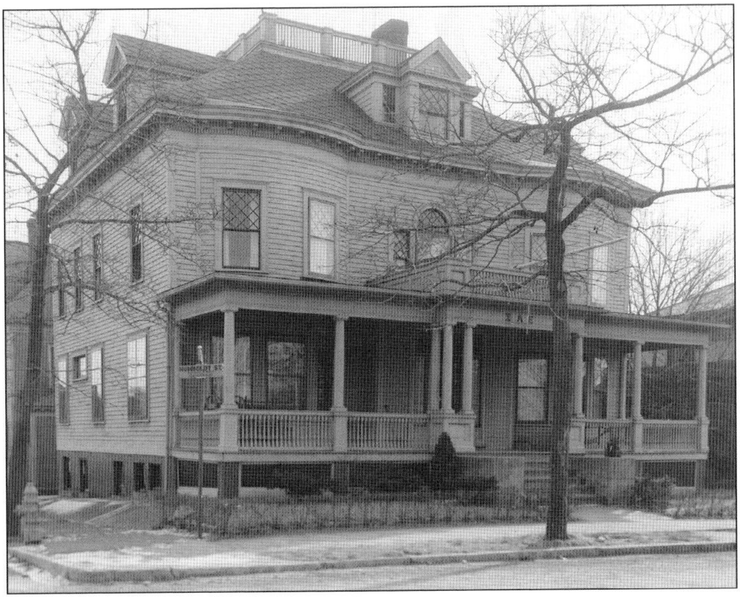

In 1905, Goddard joined the Massachusetts Delta chapter of Sigma Alpha Epsilon (SAE) fraternity. The organization provided him with a valuable social outlet and a sense of camaraderie beyond his academic pursuits. This lithograph shows what the fraternity house likely looked like during Goddard's time at WPI. While he never lived in the house, he was an active member and maintained ties with fellow brothers.

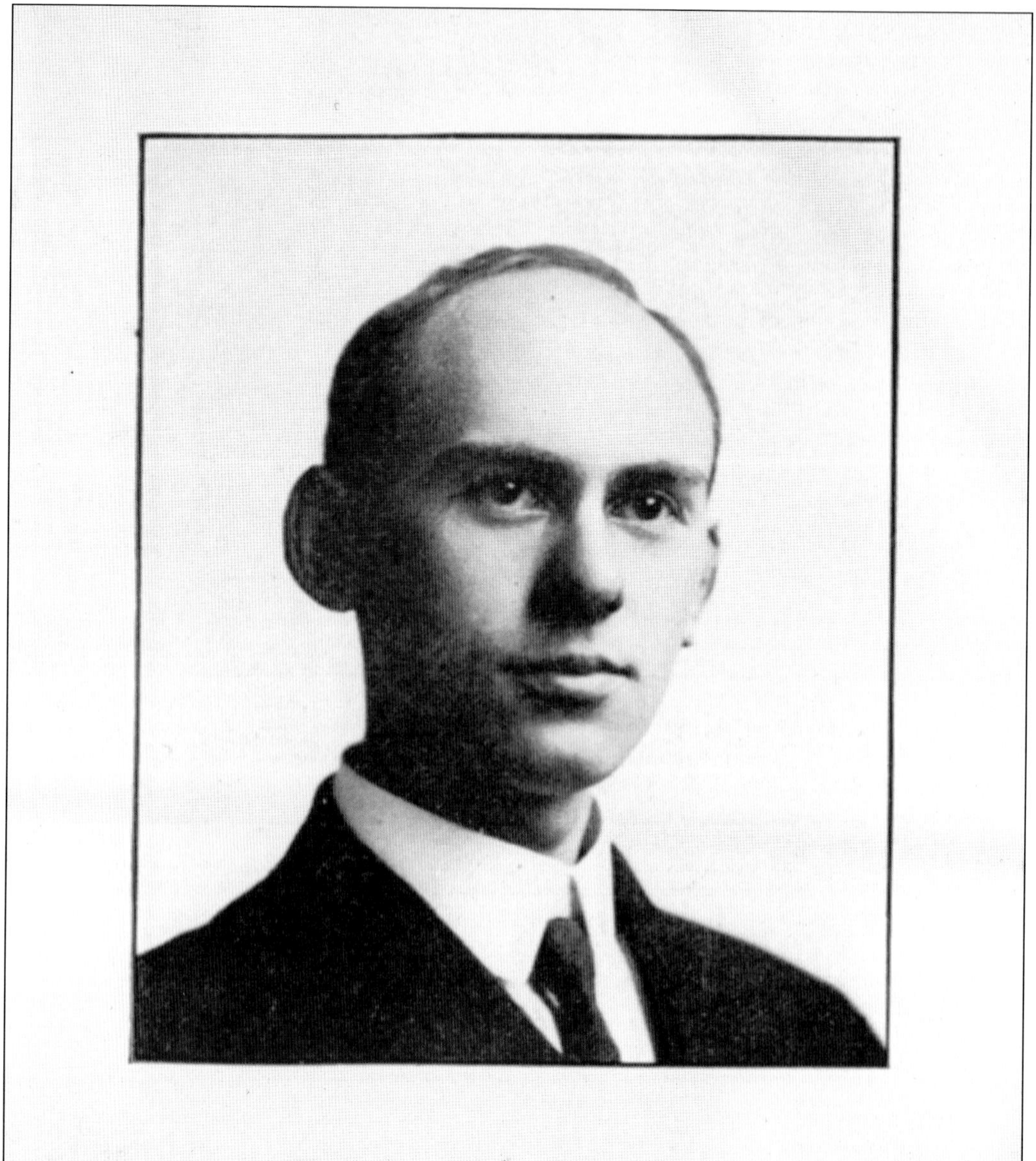

In addition to joining the Sigma Alpha Epsilon fraternity in 1905, Goddard immersed himself in a wide range of other extracurricular activities while an undergraduate at WPI, a reflection of his characteristic drive and curiosity. Much like in his high school years, he pursued leadership roles and was elected vice president of the class, followed by class president the following year. While he was quiet and reserved for much of his life, he still craved interaction with others and active participation in life. But his involvement was not limited to social or class leadership circles. He also joined Sigma Xi, the national honor society for scientific and engineering research, which recognized his early talents in experimental physics and engineering.

Goddard served as editor of WPI's yearbook, *Aftermath*, in his senior year. One line in his yearbook biography accurately states, "The word 'shark' fails to convey his appetite for knowledge." Another line teasingly observes, "On account, probably of his extreme age and the venerable air which his bald pate has given him, Bob has been continually before the class in official capacities."

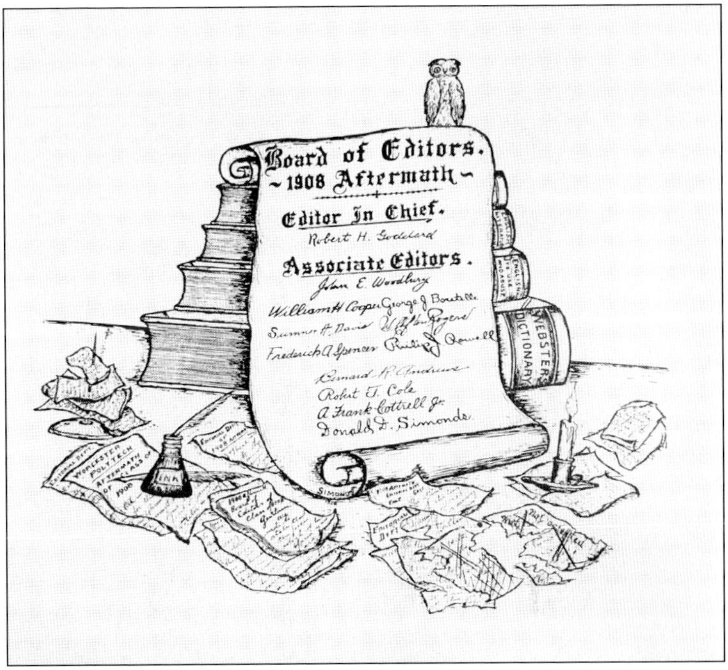

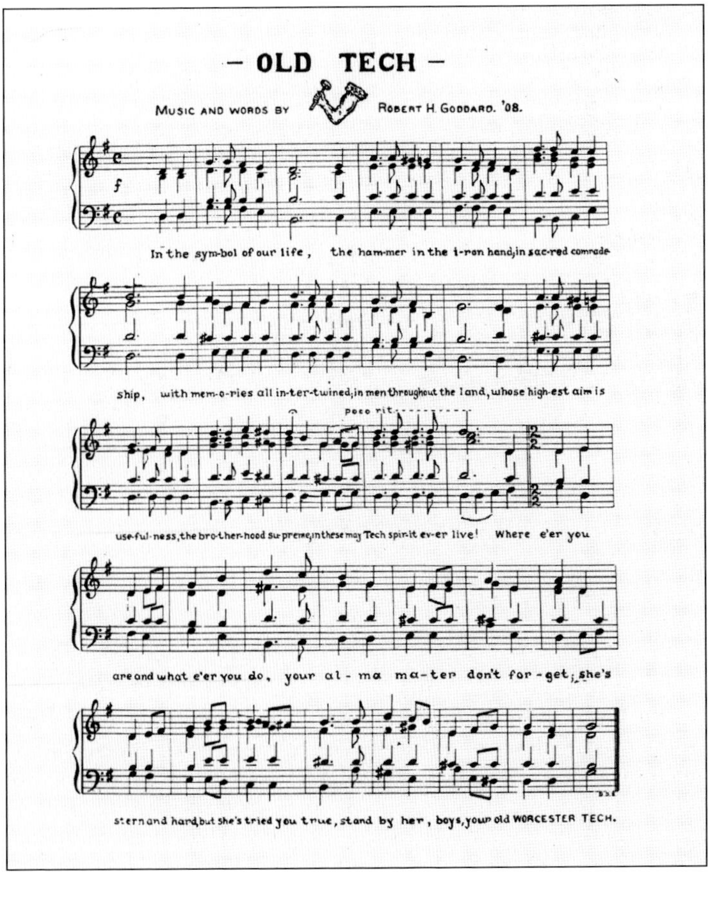

Like he had done at South High School, Goddard composed a song at WPI, titled "Old Tech." He wrote both the words and music in a way that reflected his creative and sentimental sides. Goddard maintained an interest in music throughout his life, and like other scientists before and since, he seemed to appreciate commonalities, such as mathematical structure and creativity, between music and science.

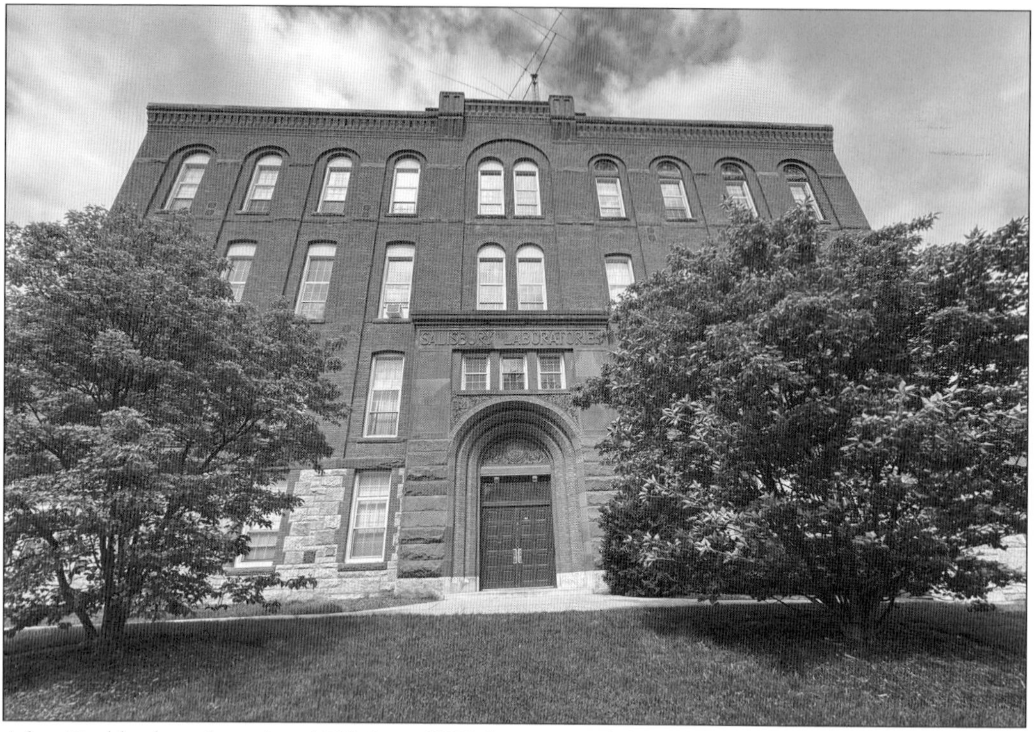

After Goddard graduated in 1908 from WPI, he returned for a short time as a staff member and continued conducting experiments on campus. He initially carried out this work in the Salisbury Labs building. As legend has it, he was forced to relocate in 1917 after an explosion that occurred while he was conducting rocket fuel research. (Kevin Schindler personal collection.)

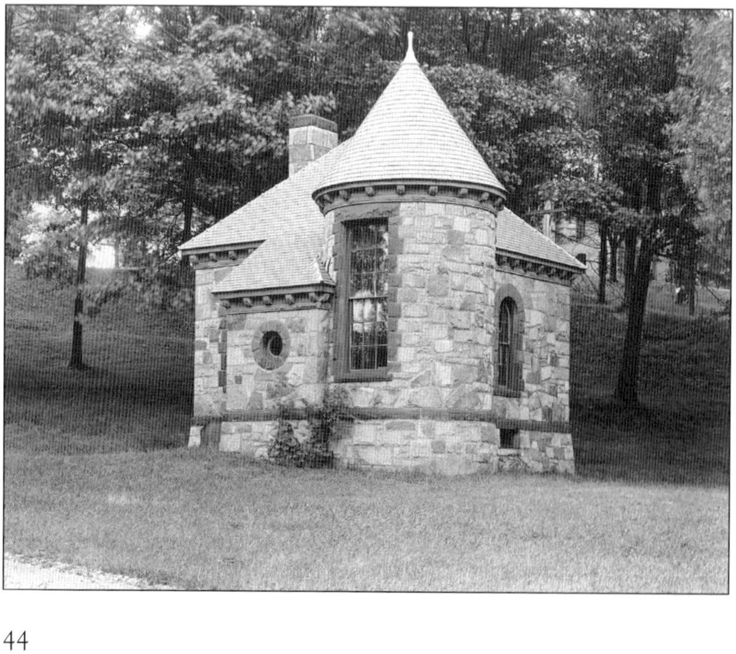

After the Salisbury Labs explosion, Goddard moved his operation to the more remote Magnetic Lab, a small rock building located on the outskirts of the WPI campus near the northwest corner of Institute Road and Boynton Street. It was built in 1887 to house an electromagnetic research laboratory, but several years later, the Skull Senior Society took over the building, and it became known as Skull Tomb.

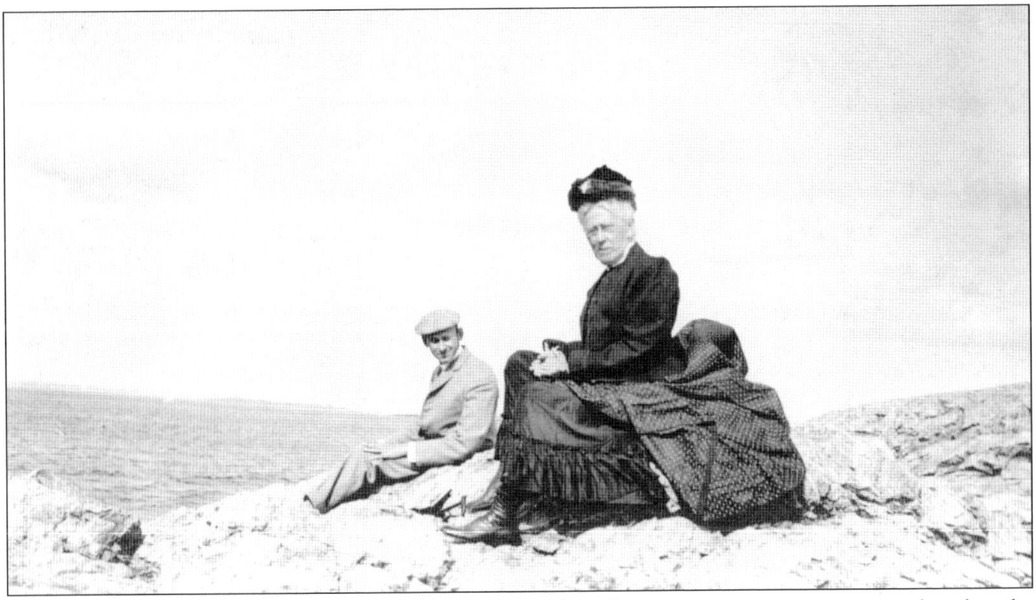

Mary Pease Upham Goddard, Robert's paternal grandmother, was a nurturing woman who played a pivotal role in his life until her death in 1916. Almost like a second mother to Robert, she cared for him during his frequent bouts of illness and even moved with him to Princeton University in 1912 to keep house for him. They are shown here during a 1908 trip to the coastal town of Marblehead.

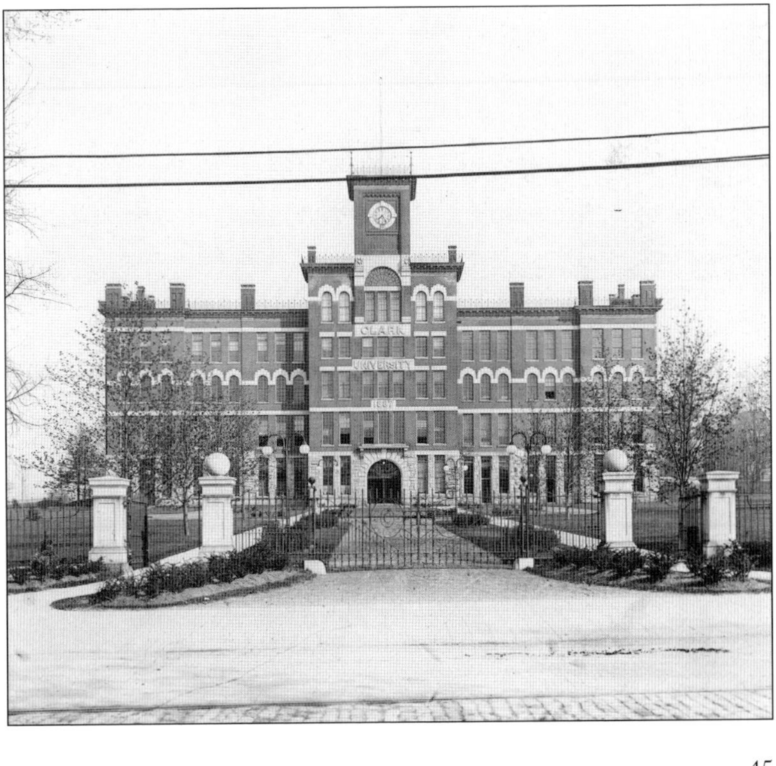

Founded in 1887, Clark University ranks among the oldest research institutions in the United States. Its campus centers around Clark Hall, pictured here in a 1890s photograph, in which it is framed by the overhead wires of Worcester's streetcar system. Goddard began his graduate studies at Clark in 1909, earning master's and doctorate degrees before embarking on a long career of research and teaching that would span decades.

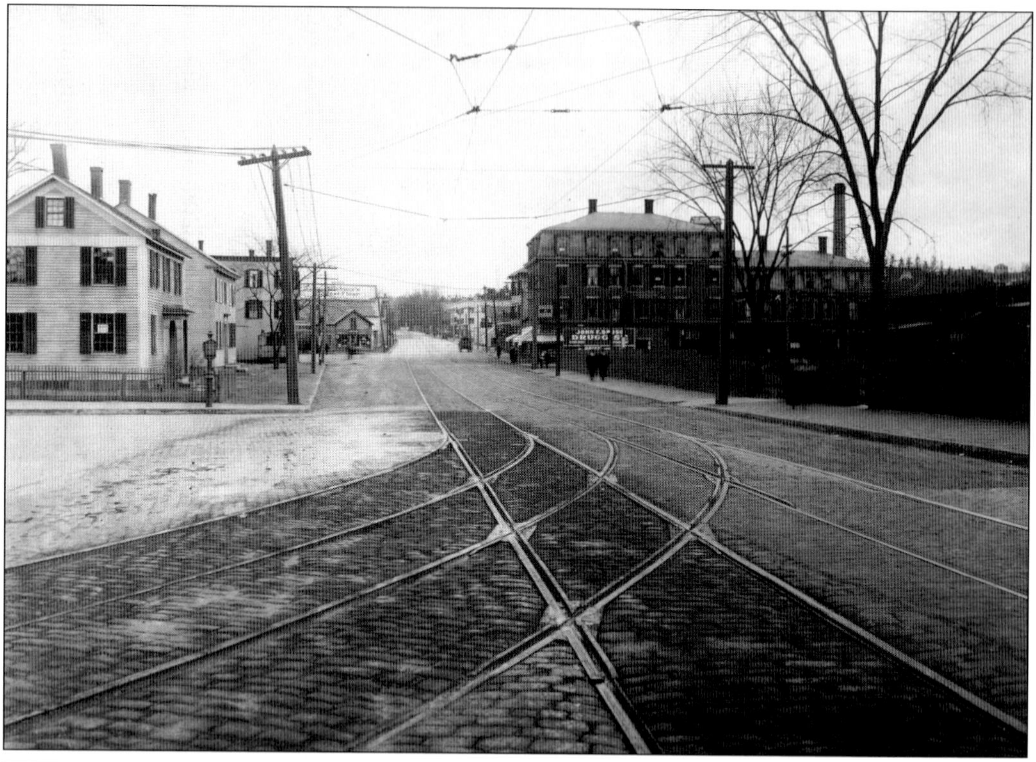

Worcester launched its first electric streetcar line in 1891, replacing the long-standing horse-drawn system. Before owning an automobile, Goddard frequently relied on these streetcars to travel between his Maple Hill home and key destinations around town, such as WPI and Clark University. He may well have ridden this very route along Main Street at Webster Square. (Museum of Worcester.)

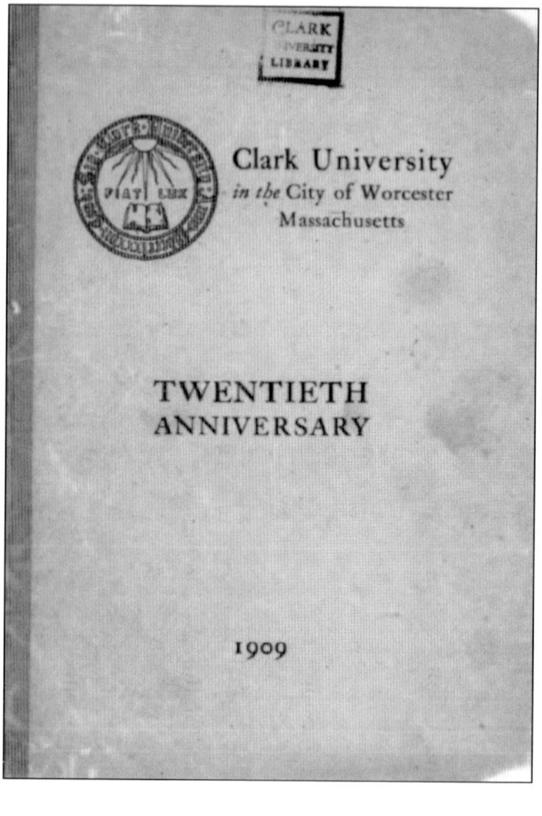

The year 1909 marked the 20th anniversary of Clark University. To celebrate, Pres. G. Stanley Hall organized a conference that brought together some of the world's leading minds in fields ranging from psychology to mathematics and physics. Goddard, then a first-year graduate student, made the most of the inspiring opportunity and immersed himself in the event, attending as many lectures and demonstrations as he could fit into his schedule. (Kevin Schindler personal collection.)

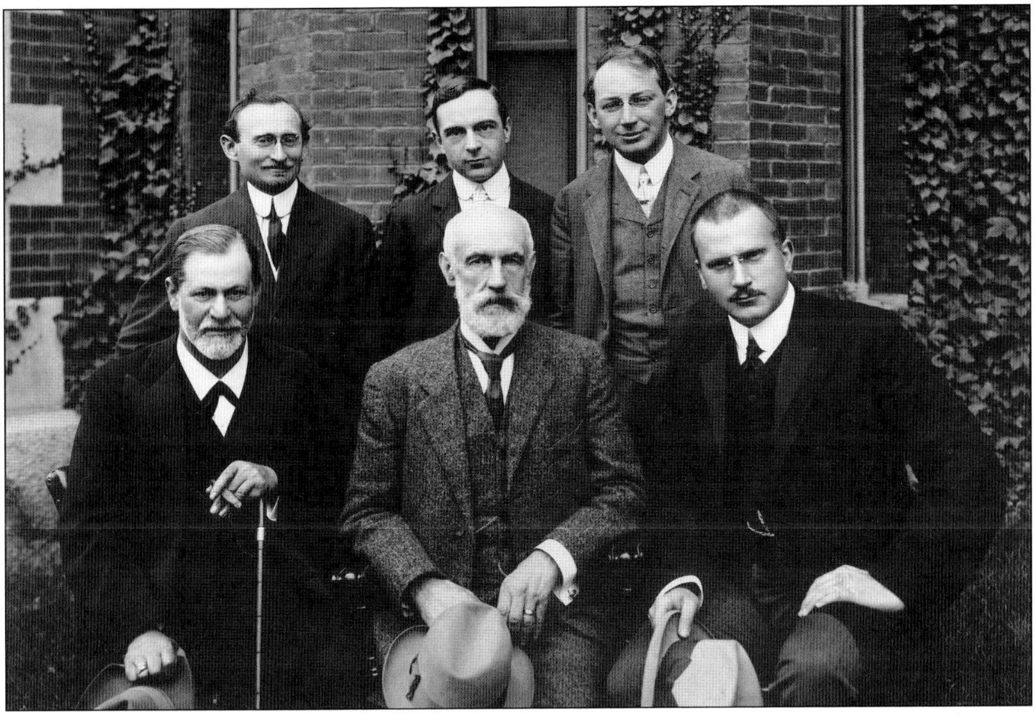

One of the most celebrated participants in the 1909 conference was the Austrian psychiatrist Sigmund Freud, who made his only trip to the United States in order to attend. A statue of Freud installed on the Clark campus in 1999 commemorates this historic visit. In a photograph (above) taken during the event, Freud is seated at left in the first row alongside Clark president G. Stanley Hall and fellow psychiatrist Carl Jung. Behind them stand Abraham Brill, Ernest Jones, and Sándor Ferenczi. Equally impressive was the gathering of scientists, which included Nobel laureates Albert Michelson and Ernest Rutherford. In a group photograph of physicists, mathematicians, and chemists (below), a balding Robert Goddard appears in the back row on the left side. Rutherford, with a moustache and wearing a dark suit, is the second person to the right of Goddard. Michelson is in a dark suit, second from right in the first row.

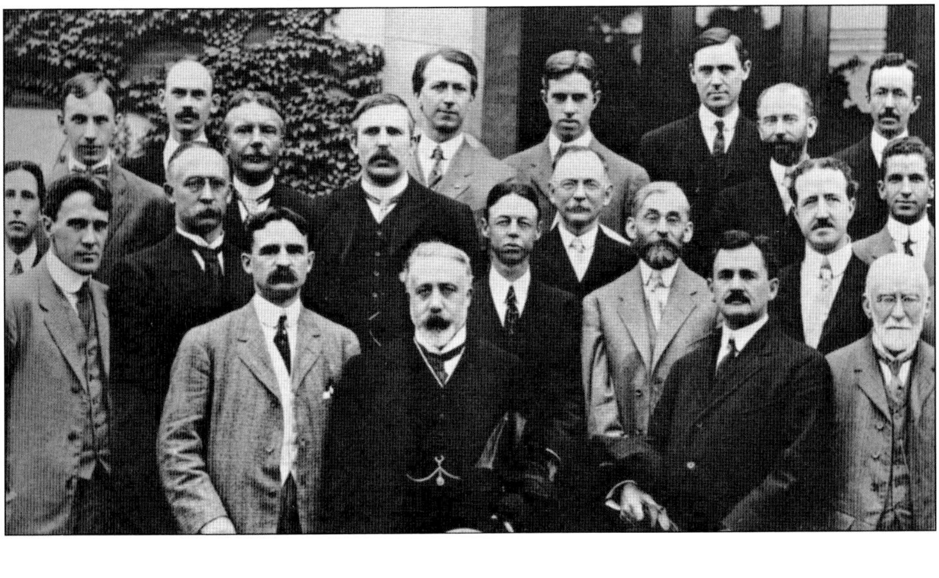

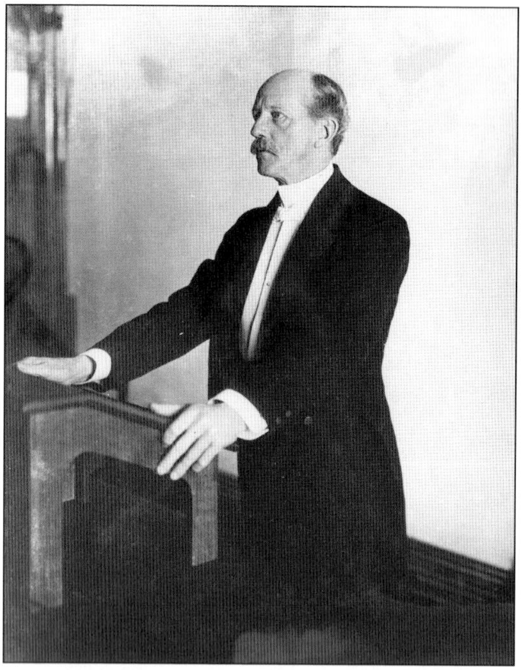

Among the distinguished participants in the 1909 Clark conference was astronomer Percival Lowell, founder of the astronomical observatory in Flagstaff, Arizona, that bears his name. Lowell gave an evening presentation about the planet Venus and also exhibited astronomical images captured at his observatory. Goddard attended Lowell's lecture and noted it in his diary. (Lowell Observatory Archives.)

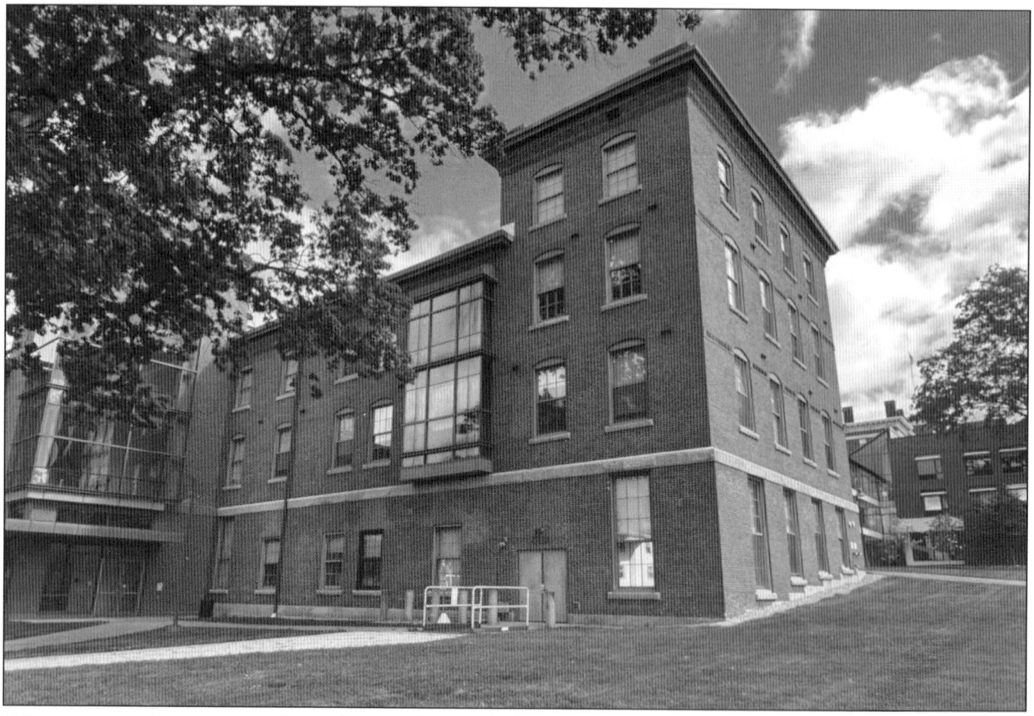

After earning his master of arts degree from Clark University in 1910, he completed his doctorate the following year. He remained there as an honorary fellow in physics before accepting a position at Princeton University in 1912. He returned to Worcester in 1913 after contracting tuberculosis. Following his recovery, Goddard resumed his work at Clark as an instructor and later as a professor, working out of the mathematics and physics building. (Kevin Schindler personal collection.)

For a decade and a half, Goddard spent much of his time in Clark's physics lab, where he conducted groundbreaking experiments in propulsion, developed and tested instruments, and explored other aspects of rocket flight. He is shown here standing by a pipe mechanism he used to study the efficiencies of rocket combustion and the possibilities of propulsion in a vacuum.

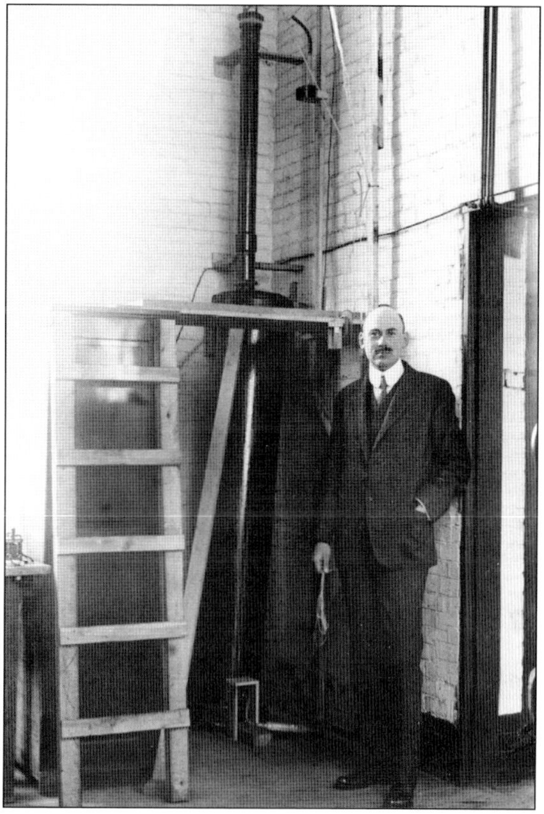

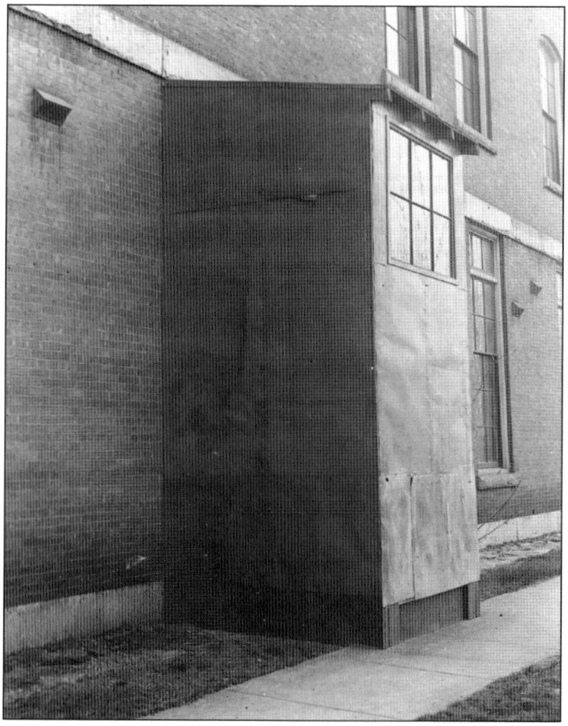

Adjacent to Clark University's physics lab stood this modest sheet-iron shed. Though unassuming in appearance, it provided Robert Goddard and his colleagues with additional space to conduct experiments, such as combustion chamber tests, that were too noisy or otherwise not ideal for the main lab. The shed was removed years ago but remains in memory as a reminder of the humble beginnings of Goddard's rocketry research.

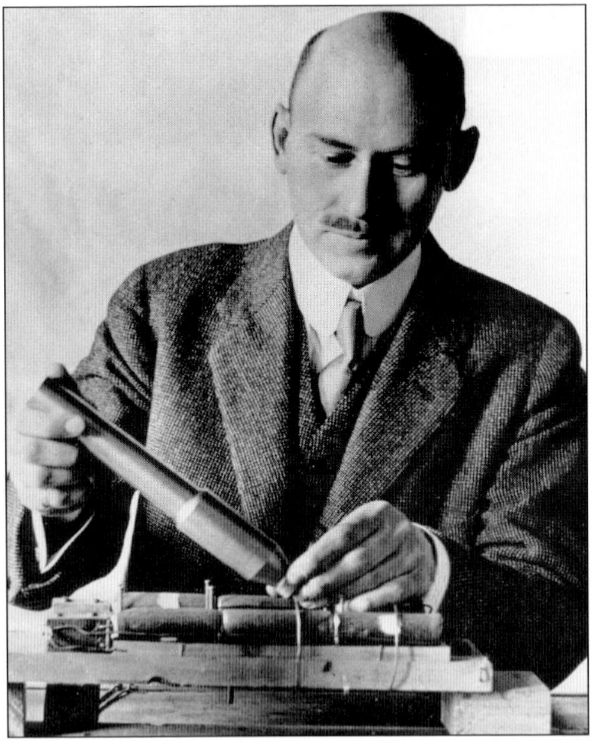

Throughout his groundbreaking career in rocket science, Goddard conducted extensive tests on every aspect of rocket design. By systematically examining and modifying the structure, shape, and composition of components such as chambers and nozzles, as shown here, he improved his understanding of how they worked and applied this knowledge to future designs.

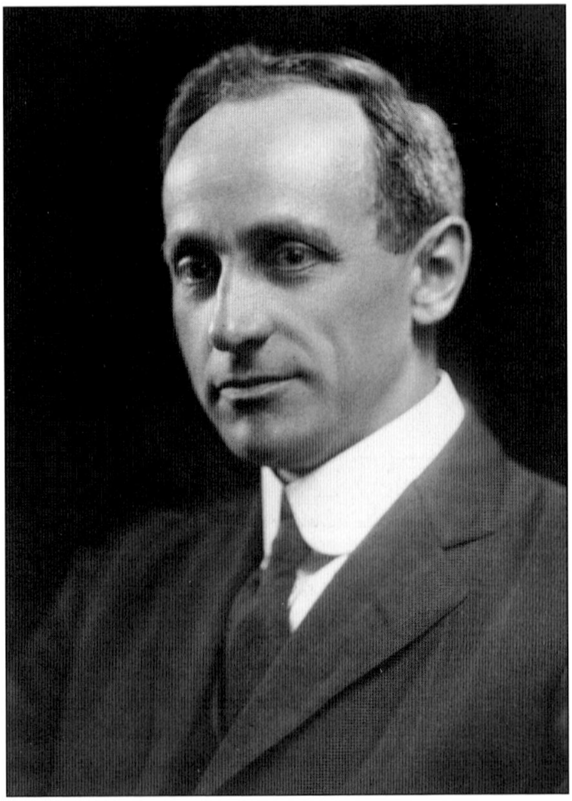

Beginning in 1914, Goddard began securing patents for the innovative devices and procedures he was developing through his rocket research. Recognizing the importance of protecting his intellectual property, he partnered with Charles Hawley, a patent attorney with the Worcester-based firm Southgate and Southgate. Hawley worked with Robert Goddard and, after Robert's death, his widow, Esther, in procuring a total of 214 patents in Robert's name.

Goddard was granted his first patent, US Patent No. 1,102,653, on July 7, 1914, for an invention simply titled "Rocket Apparatus." It recognized his priority in pioneering concepts such as the use of multi-stage rockets that Goddard realized were necessary in order to achieve high distances and velocities, a nozzle for efficiently generating thrust, and liquid fuel.

UNITED STATES PATENT OFFICE.

ROBERT H. GODDARD, OF WORCESTER, MASSACHUSETTS.

ROCKET APPARATUS.

1,102,653. Specification of Letters Patent. **Patented July 7, 1914.**

Application filed October 1, 1913. Serial No. 792,707.

To all whom it may concern:

Be it known that I, ROBERT H. GODDARD, a citizen of the United States, residing at Worcester, in the county of Worcester and State of Massachusetts, have invented a new and useful Rocket Apparatus, of which the following is a specification.

This invention relates to a rocket apparatus and particularly to a form of such apparatus adapted to transport photographic or other recording instruments to extreme heights. Certain features of the invention are also applicable to the display of signals or to the projection of explosives.

Mathematical analysis shows that in any rocket apparatus of given mass, the necessary propelling charge varies according to an expression in which the percentage of the heat energy of the charge which is transformed to kinetic energy enters in an exponential relation. Hence, any increase in the efficiency in the transformation results in greatly increased velocity of the apparatus and also permits a reduction in the amount of explosive used.

It is one of the objects of my invention to provide a rocket apparatus in which the transformation above mentioned is performed with great efficiency whereby the velocity and range of flight are greatly increased.

A second object of my invention is to provide means by which, after the apparatus as a whole has performed a given flight, the recording apparatus may be given a further flight in an auxiliary device by the use of a second but reduced propelling charge.

Further objects of my invention are to provide improved means for initially rotating the rocket apparatus; to provide means for maintaining the speed of rotation thus attained; and to provide means for so supporting the recording apparatus within the rocket that said apparatus shall not partake of the rotation of the rocket apparatus as a whole.

With these objects in view my invention comprises certain devices, arrangements and combination of parts which will be hereinafter described and more fully set forth in the appended claims.

A preferred form of my invention is shown in the drawings in which—

Figure 1 is a longitudinal view, partly in section, of my rocket apparatus as a whole: Fig. 2 is an enlarged longitudinal sectional view of the head of the apparatus: Figs. 3 and 4 are enlarged transverse sectional views taken along the lines 3—3 and 4—4 respectively, in Fig. 1: and Fig. 5 is a vertical elevation of a frame work from which the apparatus may be fired, drawn to a reduced scale.

Referring to Fig. 1, the rocket apparatus comprises a primary rocket having a casing containing a combustion chamber 10 from which depends an elongated tapered tube 11. The explosive material is indicated as a plurality of disks 12 secured within the chamber 10 by a casting 13. The disks 12 are preferably formed of a series of materials having progressively increasing rates of combustion so that as each disk is ignited it burns with increased rapidity and keeps the pressure in the chamber 10 constant, for which specific pressure the tapered tube 11 is designed. Similar results may be attained by using the explosive in a single mass of progressively varying composition in place of the plurality of disks here shown, and this construction is also within the scope of my invention.

In the ordinary forms of rocket apparatus, the gases of combustion are discharged through an opening at the rear of the apparatus but I have discovered that by providing the elongated tapered tube 11, I am able to greatly increase the efficiency of the device. The tube is shown in the form of a truncated cone of slight taper and the length of the tube should be not less than three times as great as its longest diameter, while a greater ratio than this will often be necessary to obtain satisfactory results. This construction takes advantage of the expansion of the gases as they pass through the tube and also allows the combustion of the gases to be completed before they finally issue from the tube. By experimentally determining the best proportions for the tube 11, a maximum percentage of the heat energy of the disks 12 may be transformed into kinetic energy. A fuse 14 is indicated which extends downwardly within the tube 11 and provides means by which the disks may be ignited.

In order to provide the necessary speed of rotation of the apparatus so that it may

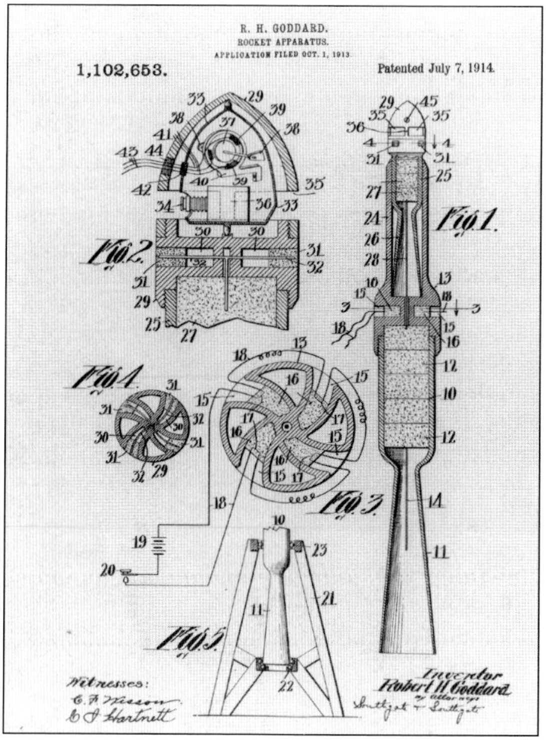

This drawing, taken from US Patent No. 1,102,653, illustrates Robert Goddard's pioneering concept for a liquid-fueled rocket. The design features two separate tanks, one for liquid oxygen and the other for gasoline. These feed into a combustion chamber and nozzle at the base, where the propellants mix and ignite to produce thrust.

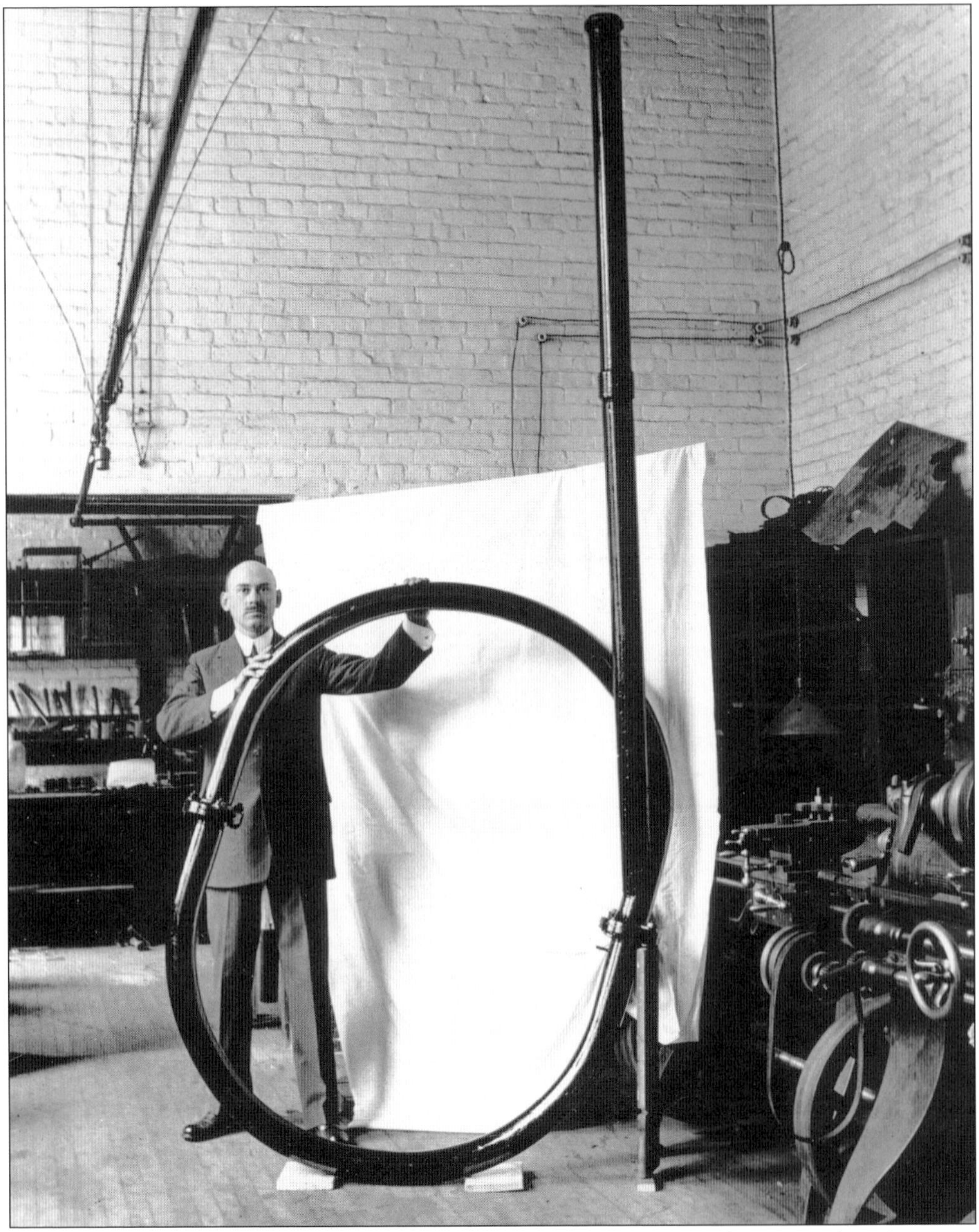

In 1916, Robert Goddard conducted a groundbreaking experiment that proved rockets could function in space. At the time, many scientists believed that, in order for a rocket to move, it needed air to push against. Thus, the thinking went, a rocket could not work in the vacuum of space. Goddard, however, did not believe this to be true. Newton's third law of motion states that for every action, there is an equal and opposite reaction. Goddard did not think the presence or absence of air should affect this, so he set out to prove his point. He placed a small rocket inside a sealed vacuum tube and expelled the air to simulate space conditions. When the rocket fired and moved within the tube, it confirmed that thrust could be generated even in the absence of air. Thrust, it turns out, is produced by the expulsion of gases, not by pushing against air. Goddard's experiment was fundamental to the development of spaceflight.

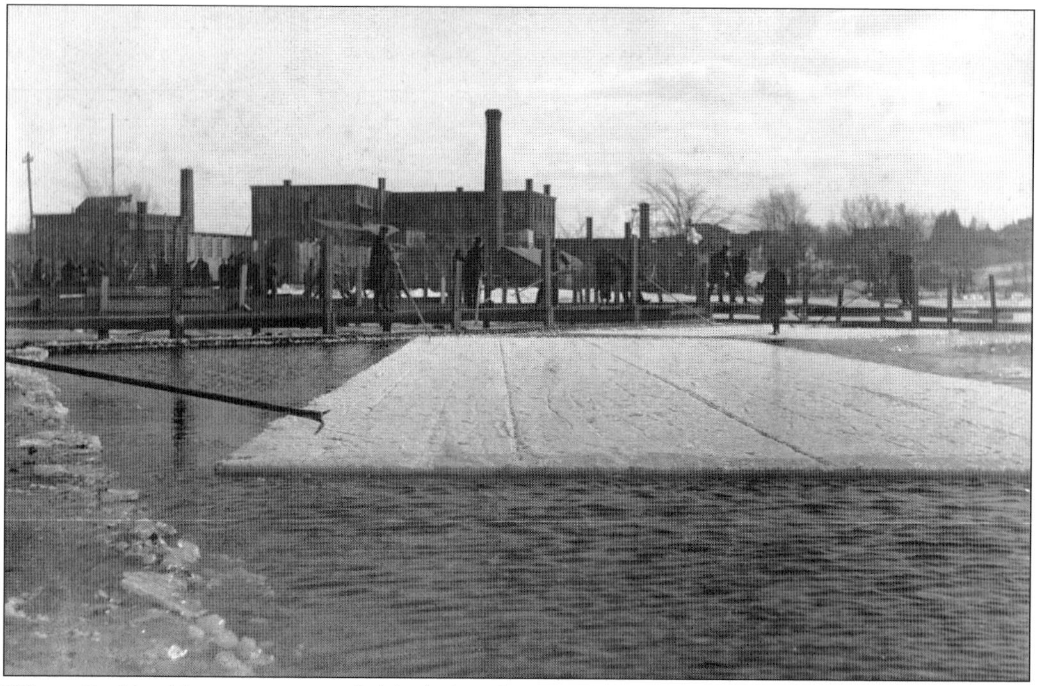

When Goddard began testing solid-fueled rockets in the mid-1910s—more than a decade before he began testing liquid-fueled versions—he waited until students had mostly vacated the Clark University campus for the night and then launched the rockets near the physics building. This drew the ire of campus workers, so Goddard then launched near Coes Pond, shown here during the winter months as ice was being harvested. (Museum of Worcester.)

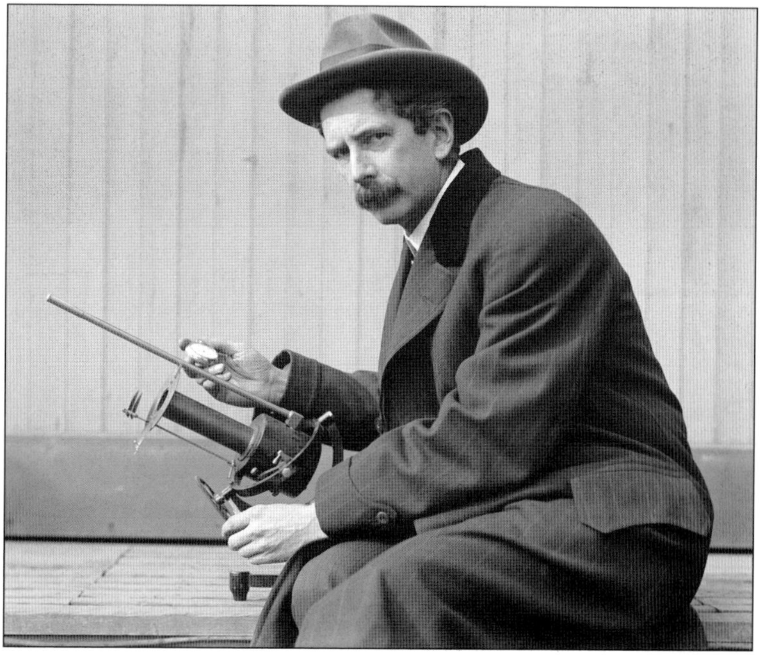

By 1916, Robert Goddard realized he needed outside funding to support his expanding rocketry experiments. He wrote a letter describing his work to the Smithsonian Institution and the following year received $5,000 on the recommendation of physicist and future Smithsonian secretary Charles Abbot. Under Abbot's direction, the Smithsonian continued to fund Goddard for several years. (Library of Congress.)

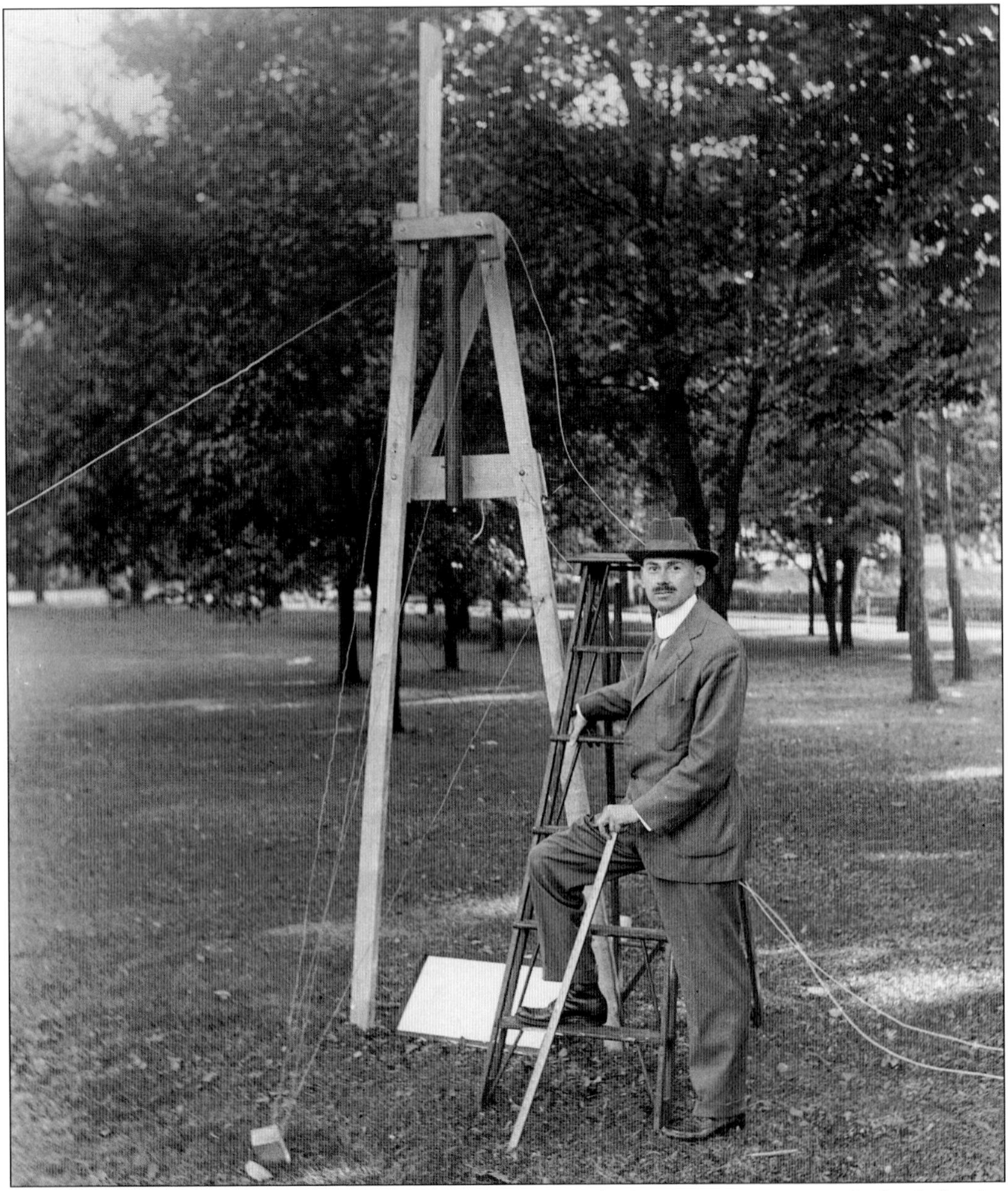

Goddard must have been quite a sight to behold on the Clark University campus for both faculty and students. While it was not unusual to see a professor in a suit strolling across the commons, watching one assemble a strange contraption on the campus lawn was something else entirely. Goddard, often dressed formally, would set up experimental devices that looked more like the whimsical inventions of a Rube Goldberg cartoon than the work of a serious physicist. And yet, these odd-looking machines were the cutting edge of rocket science. Here, Goddard stands beside the mounting for a rocket combustion chamber, one of many experimental apparatuses he designed and tested in his pursuit of spaceflight. His work combined theoretical brilliance with mechanical ingenuity, often requiring custom-built components and outdoor testing. To the casual observer, it may have seemed eccentric or even absurd, but to those who understood his vision, it was revolutionary.

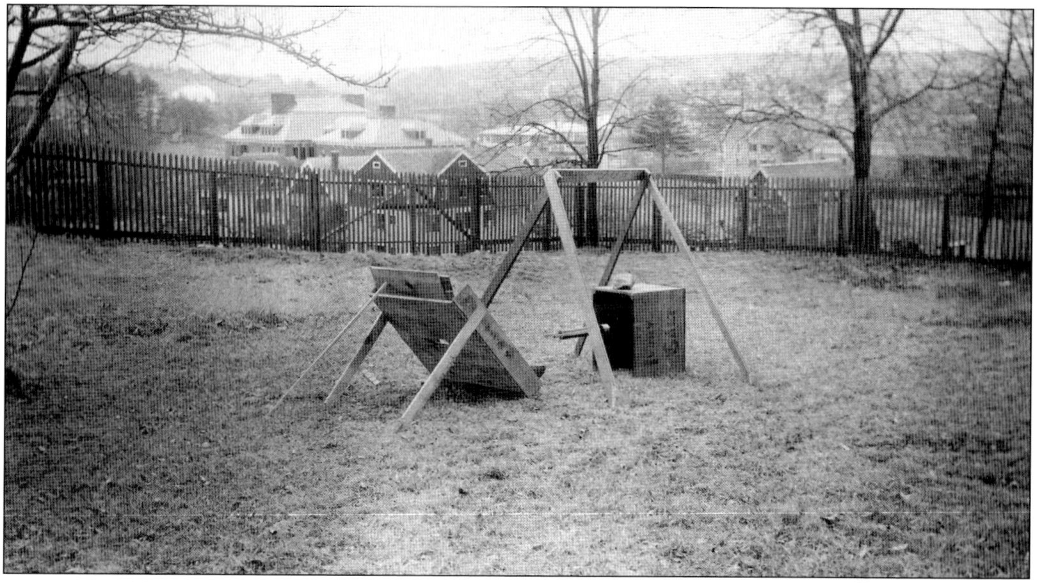

During Goddard's early years of employment at Clark University, his parents moved from Maple Hill to a new house Nahum built two streets over on Bishop Avenue. Robert continued to live at Maple Hill but, ever the opportunist, occasionally took advantage of his parents' new backyard to set up one of his experiments there.

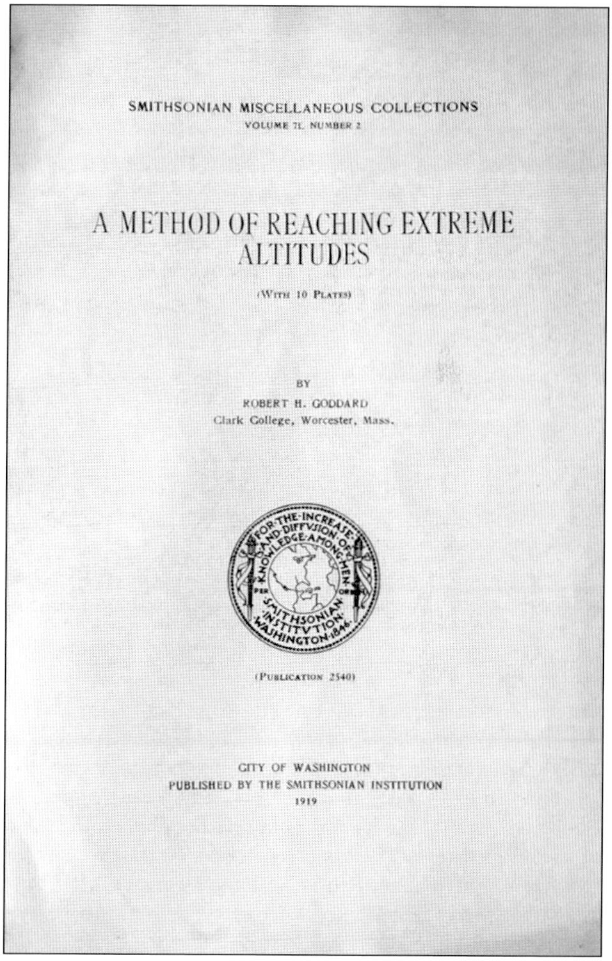

SMITHSONIAN MISCELLANEOUS COLLECTIONS
VOLUME 71, NUMBER 2

A METHOD OF REACHING EXTREME ALTITUDES

(WITH 10 PLATES)

BY
ROBERT H. GODDARD
Clark College, Worcester, Mass.

(PUBLICATION 2540)

CITY OF WASHINGTON
PUBLISHED BY THE SMITHSONIAN INSTITUTION
1919

In 1919, Robert Goddard published a groundbreaking paper that laid the foundation for modern rocket science. This 69-page document detailed the mathematical principles of rocket propulsion, described his experimental work, and proposed how rockets could be used to reach space. Published by the Smithsonian Institution, the paper was the first serious and practical guide to space travel using rockets.

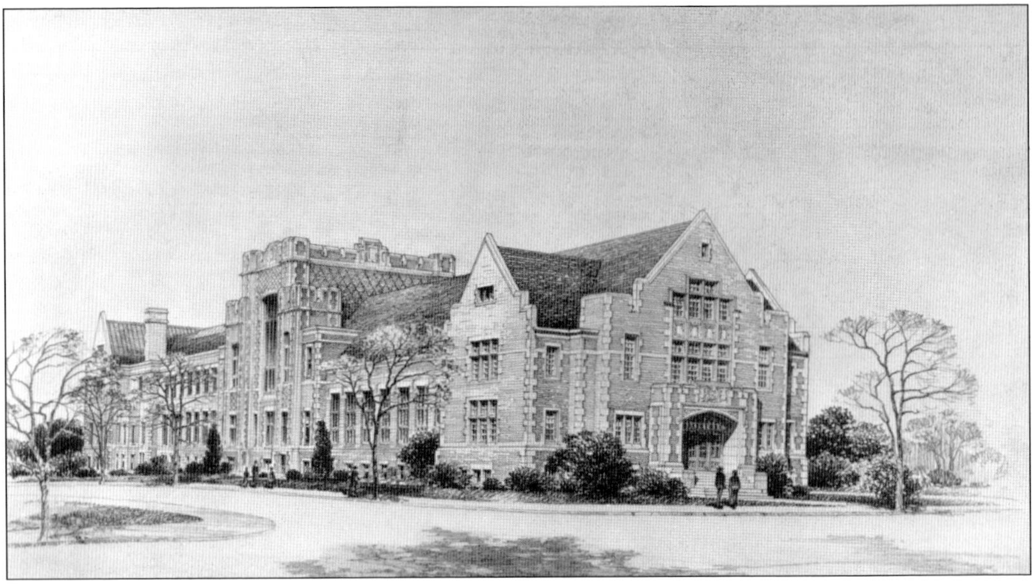

In 1920, Goddard raised money from local businessmen to purchase the transmitters, receivers, and other equipment necessary to begin radio broadcasting. He founded an AM station, WCUW, and housed it initially in Clark University's Atwood Hall (above) before moving it to the basement of Sanford Hall. WCUW broadcast a mix of music, weather forecasts, agricultural updates, and market reports. Notably, in November of its first year, the station aired live returns from the 1920 presidential election, announcing Warren Harding's victory over James Cox. This marked one of the earliest instances of real-time election coverage in the region. Though the original AM station ceased operations in 1972, it was soon revived as an FM station. Today, WCUW 91.3 FM continues to operate as a popular community radio station. (Above, Museum of Worcester; below, Robert H. Goddard Library, Clark University.)

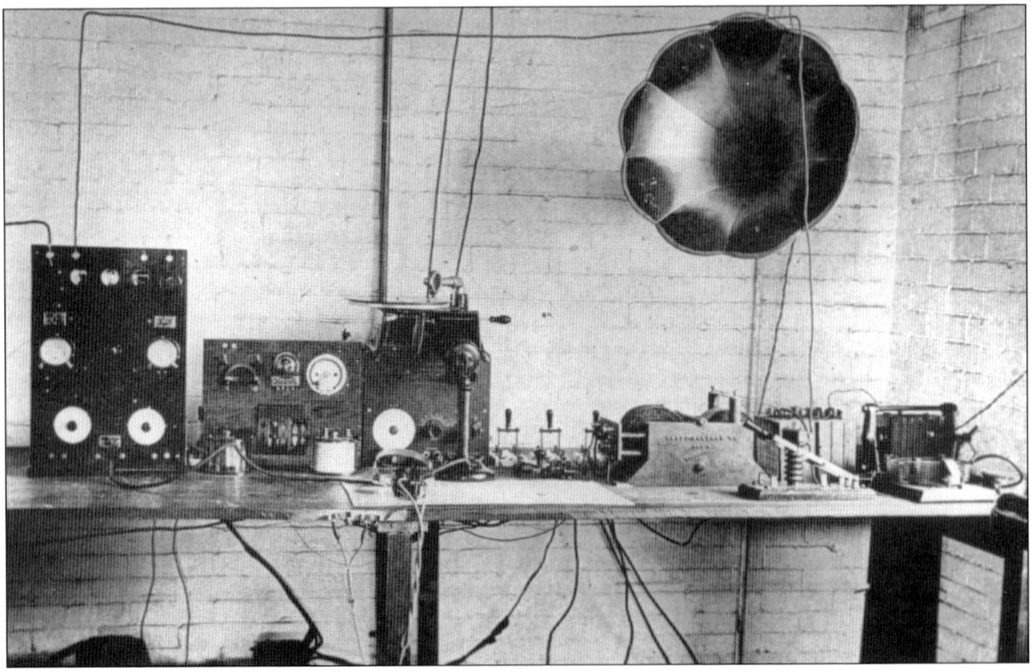

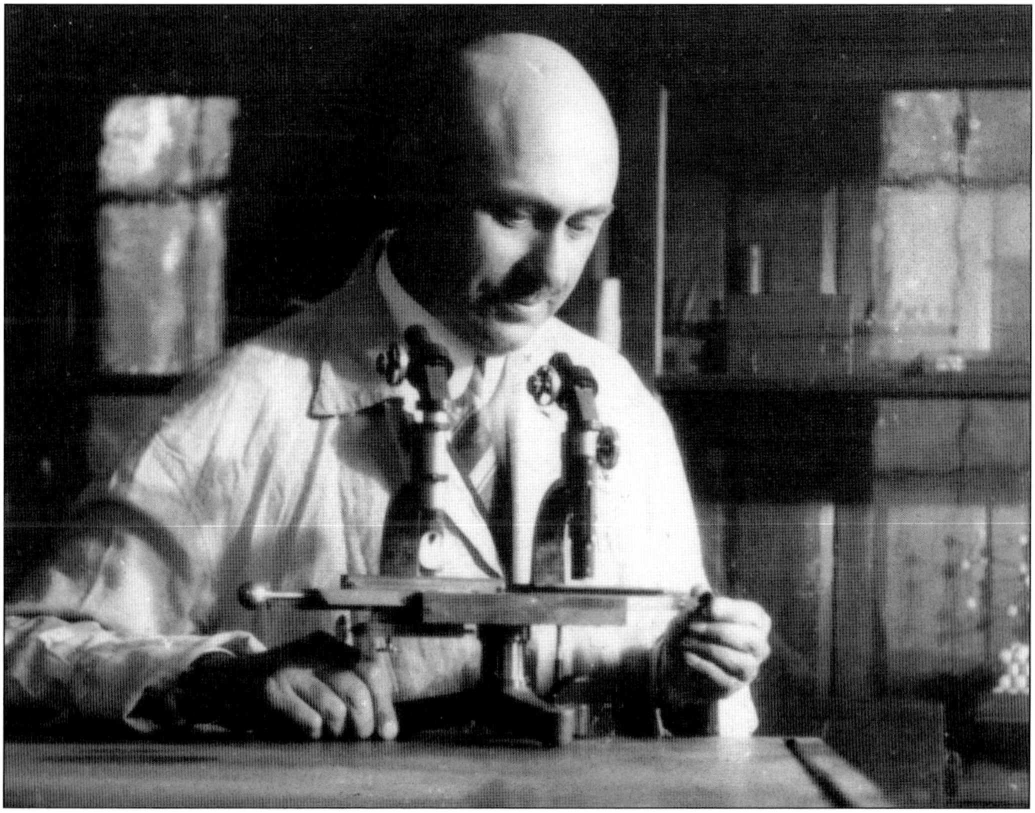

Robert Goddard's research demanded not only intellectual rigor but also practical skill with a wide range of tools and equipment. Perhaps the hands-on aspect of his work reminded him of his boyhood experiences. Goddard's father had nurtured his curiosity by providing him with microscopes, telescopes, and other scientific gear. These early gifts helped spark Robert's fascination with the natural world and laid the foundation for his lifelong passion for exploration.

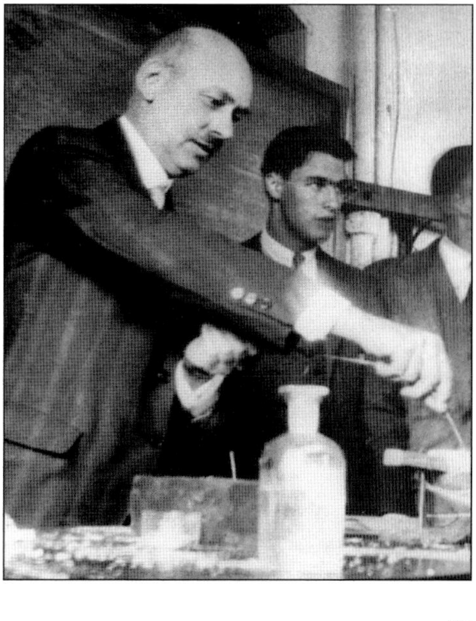

In the early 1920s, Goddard began seriously experimenting with liquid propellants as a more efficient and controllable alternative to traditional solid fuels. While he recognized that liquid hydrogen offered the highest energy potential, it was prohibitively expensive and difficult to obtain at the time. Instead, Goddard turned to a more practical combination: gasoline as fuel and liquid oxygen as the oxidizer. Luckily, Linde Company was able to provide him with liquid oxygen.

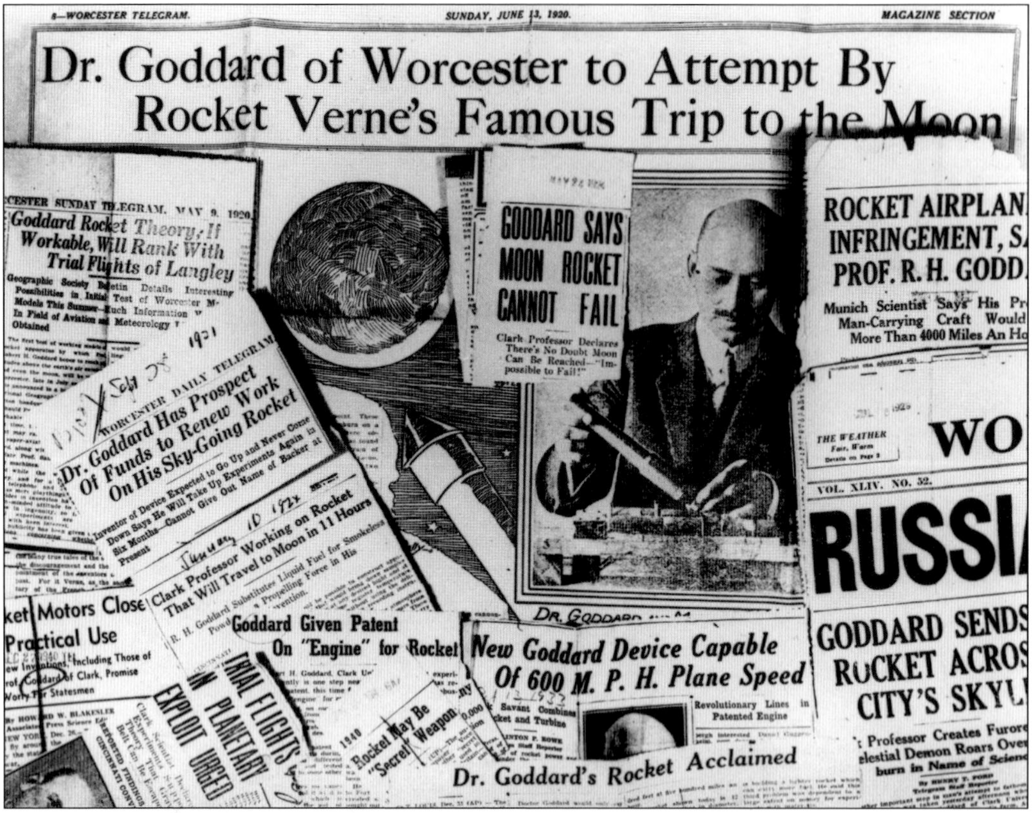

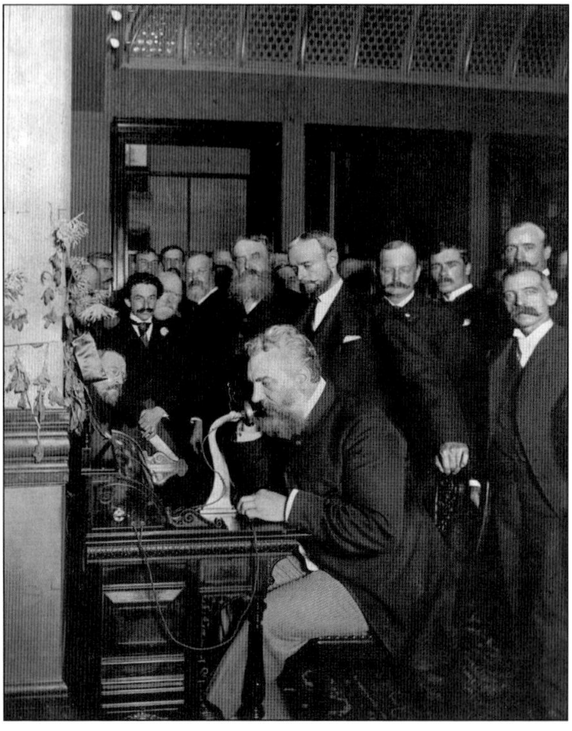

While Robert Goddard spent much of his early career outside the public spotlight, that changed dramatically with the publication of *A Method of Reaching Extreme Altitudes*. Especially provocative to the media was Goddard's modest assertion that rockets could eventually be used to travel to the Moon. They sensationalized his message to imply that he was actively attempting to launch a rocket to the Moon at that moment.

This photograph shows Alexander Graham Bell in his younger years. By 1920, nearing the end of his life, he hosted weekly "conversaziones" to discuss new scientific ideas. Introduced by the Smithsonian's Charles Abbot, young rocket scientist Robert Goddard joined one evening, sharing his visionary work. The meeting bridged generations—two pioneers united by curiosity and a belief in the transformative power of innovation.

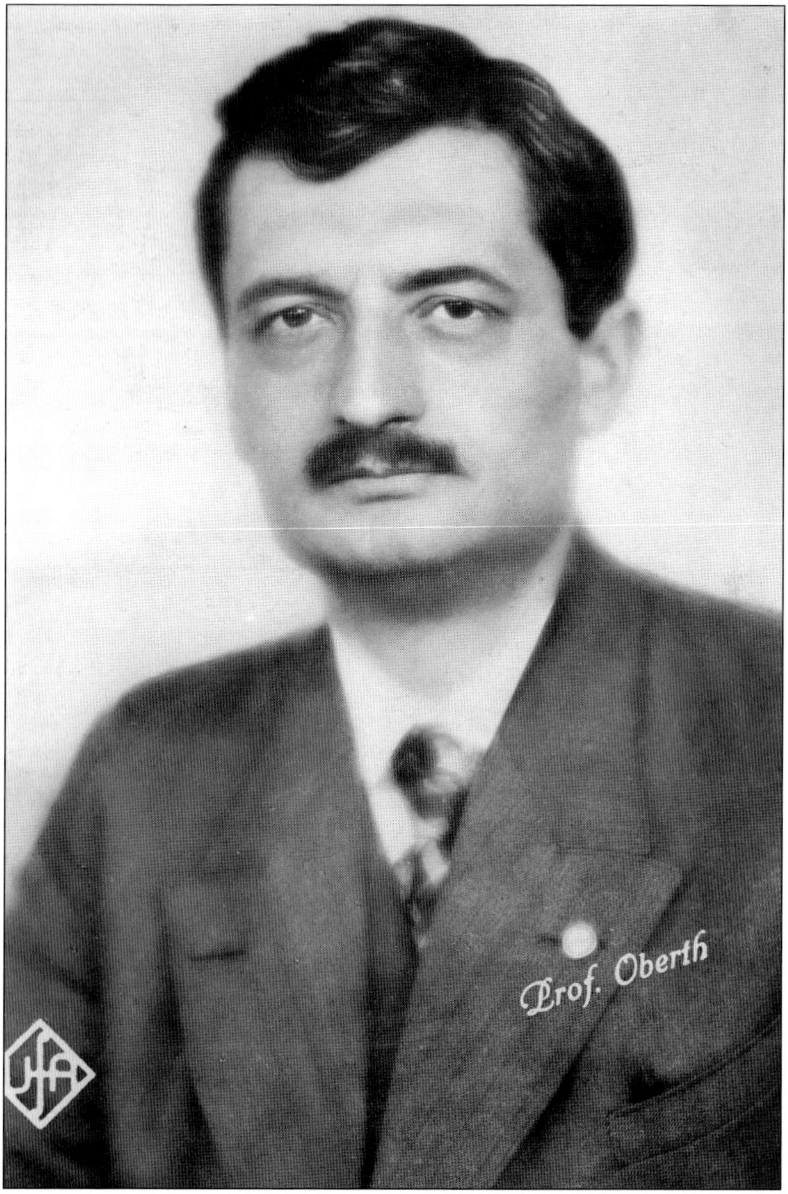

By the early 1920s, Robert Goddard's groundbreaking work in rocketry was gaining recognition, particularly among scientists intrigued by the possibility of space travel. His *A Method of Reaching Extreme Altitudes* paper had circulated internationally, sparking interest in his theories and experiments. In 1922, Goddard received a letter from a young German physicist named Hermann Oberth (shown here), who had been inspired by Goddard's work and was developing his own theoretical models for rocket propulsion. Oberth, like Goddard, believed that spaceflight was not just science fiction but a real, achievable goal. While they never collaborated directly, Oberth's influence on later German rocketry and Goddard's experimental achievements in the United States would shape the foundation of modern space exploration. Yet in considering the early development of space travel, a third person must also be recognized, a Russian theoretician named Konstantin Tsiolkovsky, who, starting in the 1880s, had written about rocketry and possible space travel. Working remotely and quietly on his own, his ideas would not be known to other scientists for years.

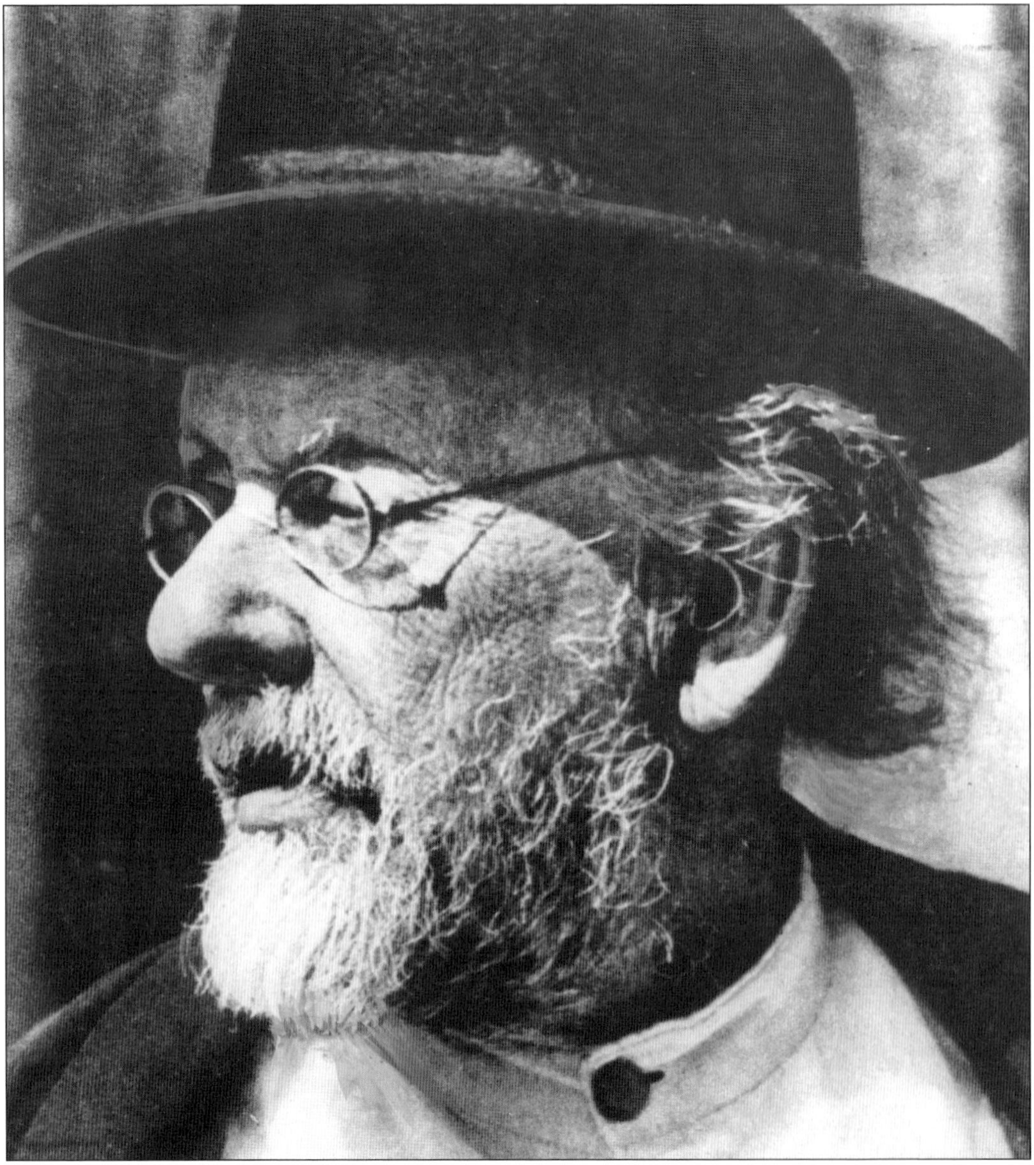

Scientist and visionary Konstantin Tsiolkovsky was born in 1857. He had little formal education and worked in relative obscurity but, inspired by the belief that humans are destined to travel into space, went on to develop groundbreaking theories about spaceflight. His work was largely theoretical—contrary to Goddard, who both theorized and tested—but his contributions were still prodigious. He envisioned liquid fuels as rocket propellants and multi-stage boosters—ideas that Goddard later independently developed and ultimately tested—as well as space stations and other concepts that later generations physically developed and tested. He famously captured his vision in a quote: "Earth is the cradle of humanity, but one cannot live in the cradle forever." Together with the contributions of Goddard and Oberth, Tsiolkovsky's work forms the foundation of spaceflight, a blend of visionary theory and hands-on experimentation with a good mix of inspiration. Their combined legacies propelled humanity toward the heavens.

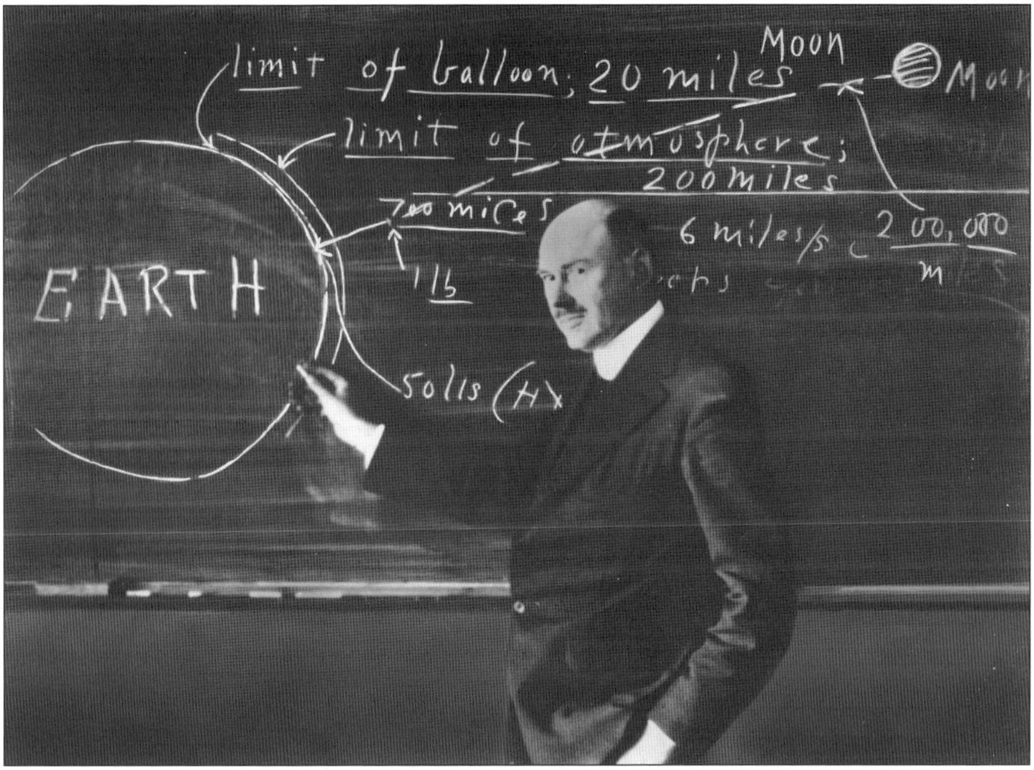

In this iconic picture taken in 1924, Robert Goddard stands at a chalkboard in a Clark University physics classroom, diagramming Earth and the Moon to explain how a rocket could reach lunar orbit. At the time, Goddard was publicly criticized—including in a disparaging commentary by the *New York Times*—about his ideas of traveling to the Moon. He remained steadfast in his convictions and was ultimately vindicated.

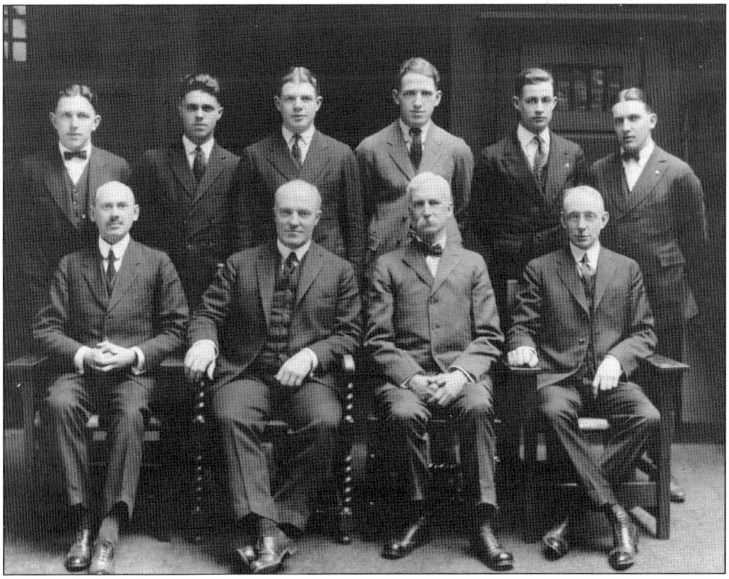

Gryphon Society, founded in 1909, was a men's senior leadership group at Clark University whose members were selected based on academic achievement and service to the community. Each year, the students invited a number of honorary faculty members to join the group. Goddard, shown here seated to the left, was so honored in 1922. The society later merged with the women's senior honor society, Pleiades.

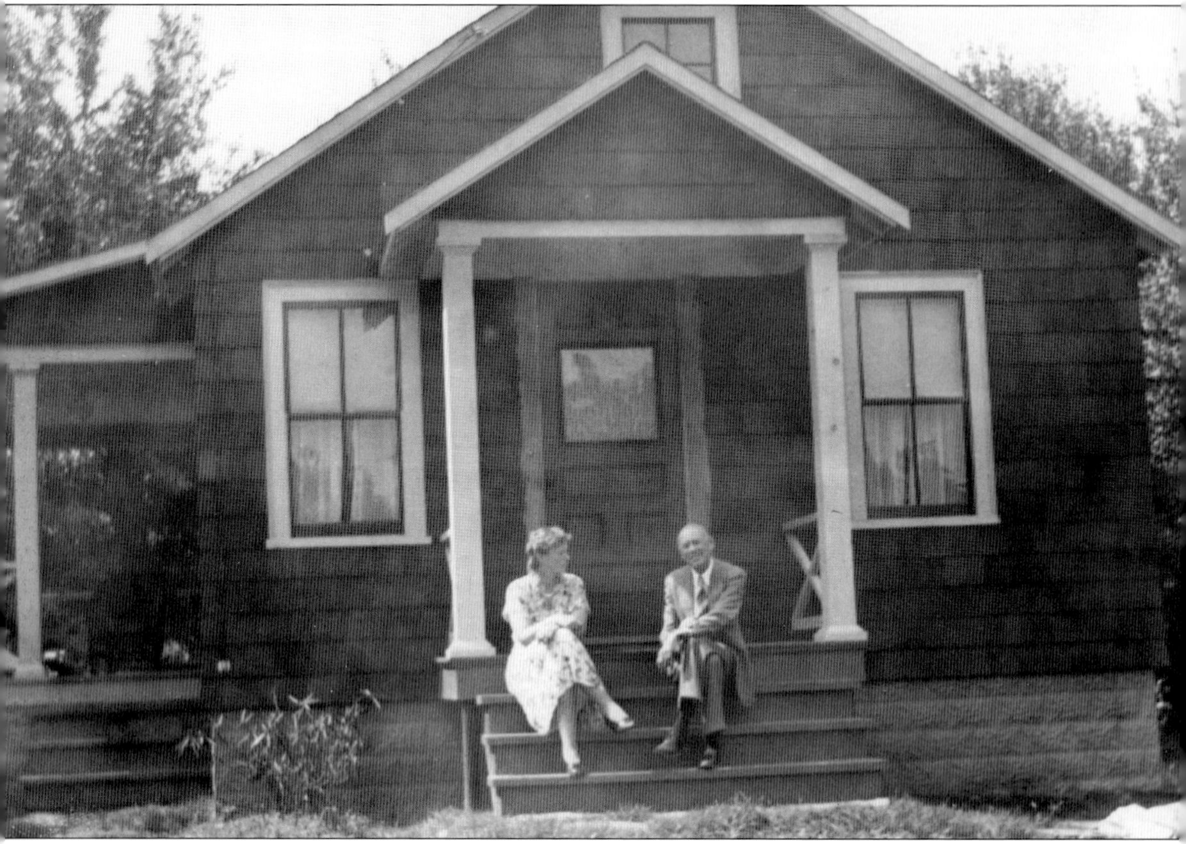

Robert Goddard's most devoted supporter—from his early years at Clark University through the end of his life—was his wife and lifelong partner, Esther Christine Kisk Goddard. Born in 1901, Esther was working as a secretary in the office of Clark's president when she met Goddard in 1919. Despite her being nearly 19 years his junior, the two quickly formed a deep bond and became inseparable. Their relationship evolved into a true partnership: while Robert's work determined where they lived, Esther played an active role in his research while maintaining their home and social life, allowing Robert to focus on his groundbreaking research. Their shared dedication to science and each other defined their marriage. This photograph captures the couple later in life during a visit to Esther's family home in Paxton, Massachusetts, a quiet moment that reflects the strength of their enduring connection.

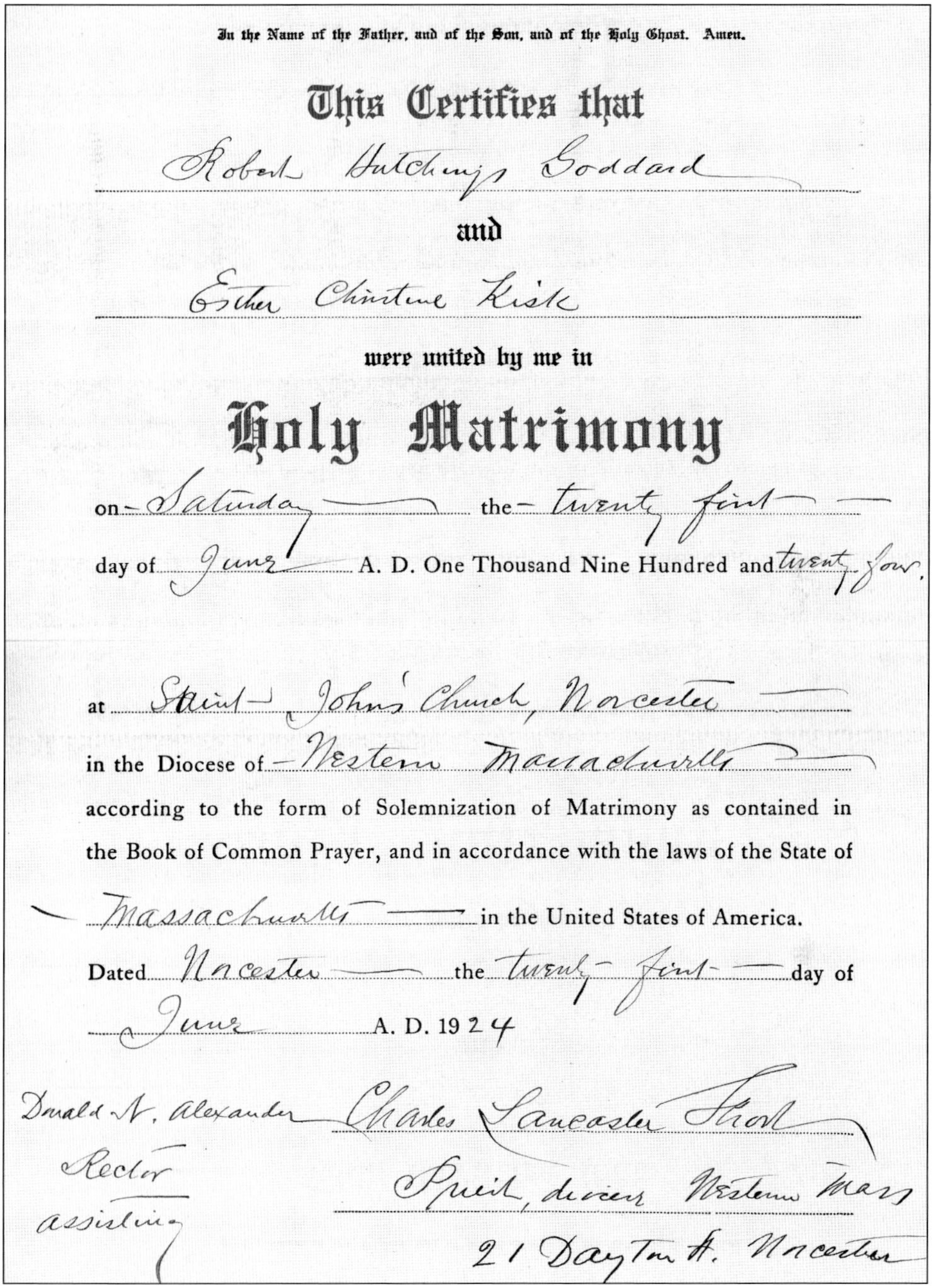
In the Name of the Father, and of the Son, and of the Holy Ghost. Amen.

This Certifies that

Robert Hutchings Goddard

and

Esther Christine Kisk

were united by me in

Holy Matrimony

on Saturday the twenty first

day of June A. D. One Thousand Nine Hundred and twenty four.

at Saint Johns' Church, Worcester

in the Diocese of Western Massachusetts

according to the form of Solemnization of Matrimony as contained in the Book of Common Prayer, and in accordance with the laws of the State of

Massachusetts in the United States of America.

Dated Worcester the twenty first day of

June A. D. 1924

Donald N. Alexander Rector assisting

Charles Lancaster Short

Priest, diocese Western Mass

21 Dayton St. Worcester

Robert and Esther Goddard were married on June 24, 1924, at St. John's Episcopal Church on Holden Street in Worcester. The church, often mistaken for the more prominent St. John's Catholic Church on Temple Street, operated from 1884 until its closure in 2009 due to financial strain and declining attendance. Rev. Donald N. Alexander, who officiated the ceremony, served as rector of St. John's Episcopal from 1918 to 1946. Though Robert had been active in church life during his youth—attending services and participating in religious activities—his adult years were largely consumed by scientific work, making his wedding one of the few occasions he returned to church. The ceremony marked a rare personal moment in a life otherwise devoted to research.

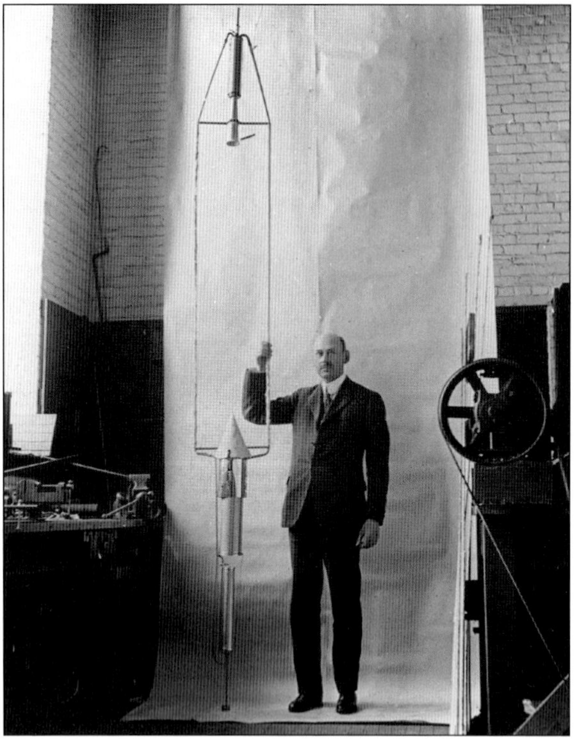

For several years, Robert Goddard experimented with liquid-fueled rocket designs and conducted static tests by attaching engines to stands so they could fire without launching. By March 1926, he felt ready to finally attempt a full launch. In this photograph taken shortly before the event, Goddard stands beside his rocket in the Clark University physics lab, preparing for a moment that would change history.

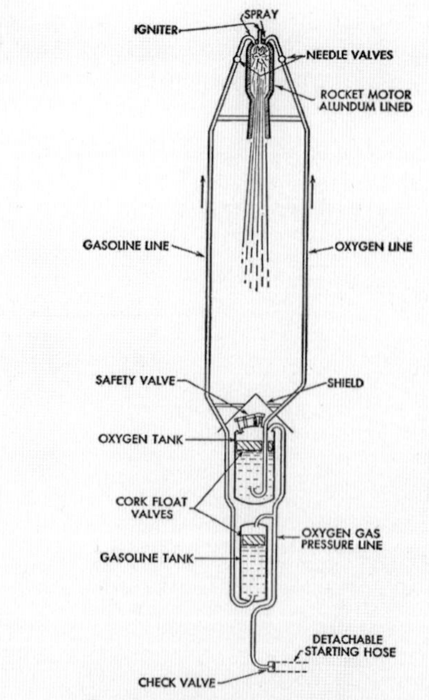

Fig. 15. Dr. Goddard's historic 1926 liquid-fuel rocket in cross section. Drawing by Dr. Goddard especially for this book.

Goddard's first liquid-fueled rocket used gasoline as fuel and liquid oxygen as oxidizer. He placed the engine above the fuel and oxidizer tanks, believing this configuration would enhance stability during flight. However, he discovered that placing the engine below the tanks did not significantly reduce stability, and it also simplified the overall design. As a result, later models featured the engine mounted below, a layout that became standard in rocket engineering.

Three

Auburn

Launching a Dream

Aunt Effie's truck farm was the Kitty Hawk of modern rocketry,
and Robert H. Goddard was its Wright Brothers.

—David Clary, *Rocket Man*, 2003

As Robert Goddard's rocketry experiments expanded, he needed a new location from which to test various components and ultimately launch the rockets. The campus of Clark University, under the cover of night, simply would not work, and neither would his backyard. However, he found a good spot courtesy of Aunt Effie Ward, a distant relative who owned a truck farm in nearby Auburn. The site was located only a few miles from Goddard's home and consisted of wide-open spaces among the cabbages and other vegetables growing there.

One of Goddard's earliest experiments at Aunt Effie's, in 1916, involved friends who ignited several different flash powders at the farm at predetermined times. Goddard, in the backyard of his home, observed the flashes and used a stopwatch to measure the amount of time that passed between when the powders were ignited and when he could see the resulting flashes from his location.

As Goddard transitioned from solid-fueled to liquid-fueled rocket experiments, he spent more time at Aunt Effie's. Then came the glorious day in March 1926 when he launched a liquid-fueled rocket. The rocket's flight was modest, rising 41 feet and staying in the air for less than three seconds. Yet it marked a revolutionary moment in aerospace history.

In addition to many rocket component tests he conducted at Aunt Effie's farm, Goddard successfully launched four rockets: the initial flight on March 16, 1926; a follow-up flight on April 3, 1926; the Hoopskirt Rocket, as seen on the cover of this book, on December 26, 1928; and the first science payload, on July 17, 1929. The last one ignited a field fire, which brought some negative attention to Goddard, but the widely reported flight also captured the attention of aviator Charles Lindbergh, who would soon befriend Goddard and become one of his most ardent supporters.

The site of Aunt Effie's farm was later converted into the Pakachoag Golf Course, where multiple historic markers pay homage to Goddard's pioneering research. One of these is an obelisk on the fairway of the ninth hole, a close approximation of the site of Goddard's first launch, where humanity took its first steps in leaving Earth and exploring the universe.

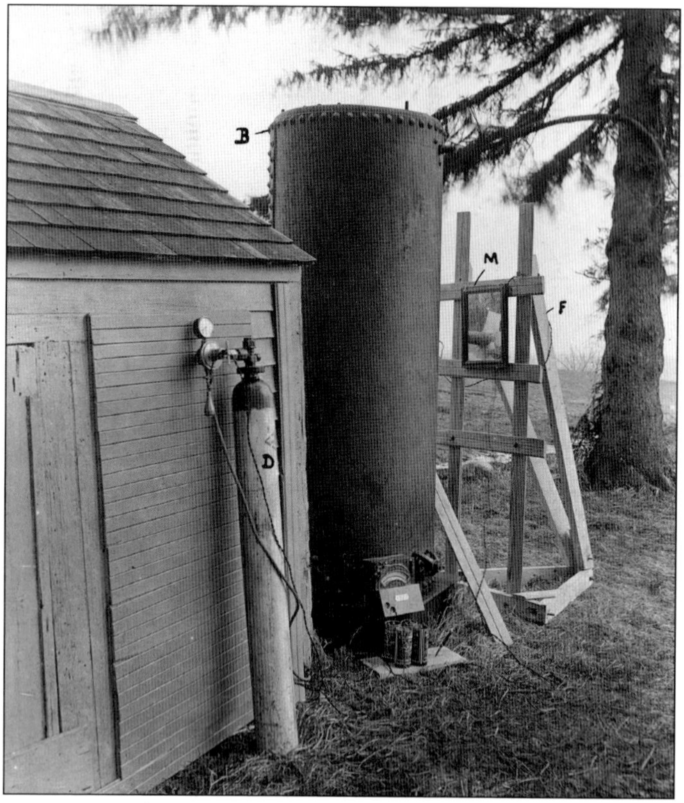

Before Goddard could attempt launching liquid-fueled rockets, he needed to conduct extensive experiments to determine key design parameters such as the optimal size, shape, and internal pressure for combustion chambers. He also had to prove that liquid propellants could generate sufficient thrust. In 1922, he used a complex apparatus of chambers, tubes, valves, and batteries to test these ideas.

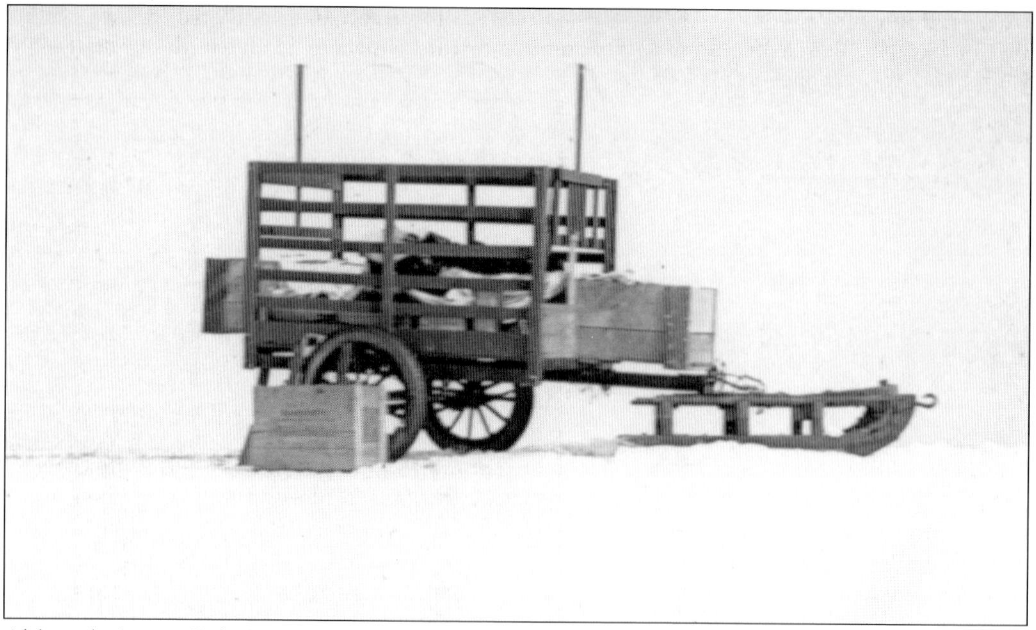

Although Aunt Effie's farm was only a few miles from Robert Goddard's home, the trip was not always an easy one. Snowy conditions, for instance, required Goddard to use a sled to transport equipment once he arrived at the farm. Here is a trailer Goddard used to haul the sled, rockets, tools, and other equipment necessary to conduct his experiments.

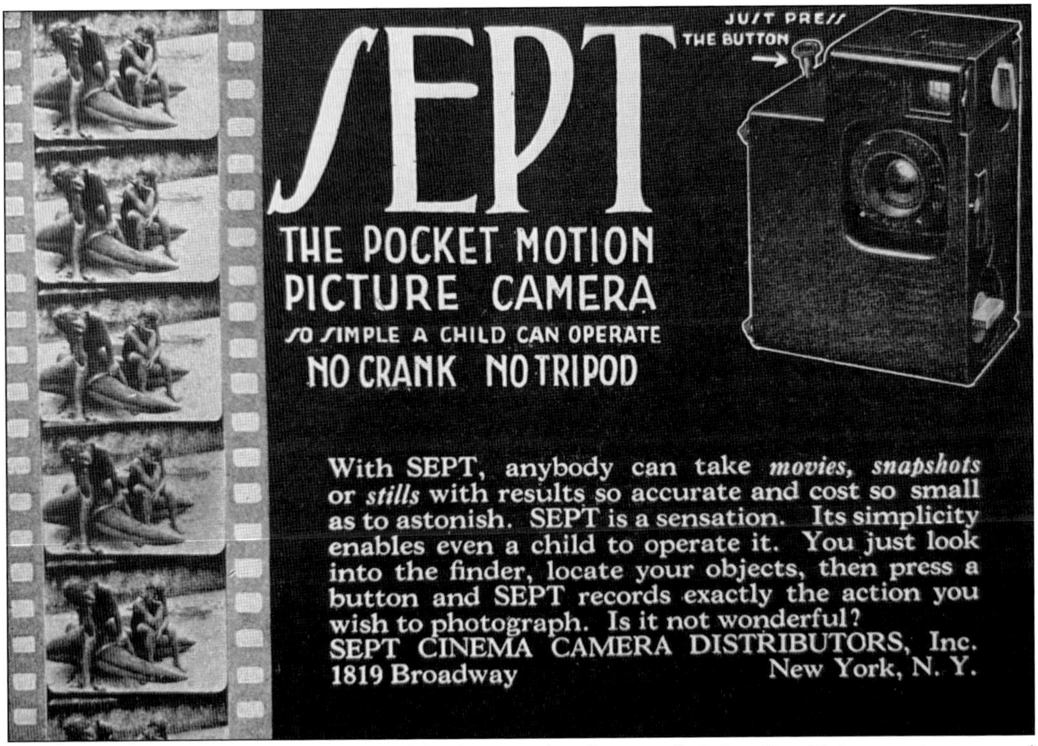

To the delight of later historians, Goddard meticulously recorded details of his experiments and rocket launches. Alongside taking extensive notes, his own personal photographer—Esther—captured images and film of key moments. For the historic March 16, 1926, launch of the world's first liquid-fueled rocket, she used a 35-millimeter Sept motion picture camera to document the event. (Kevin Schindler personal collection.)

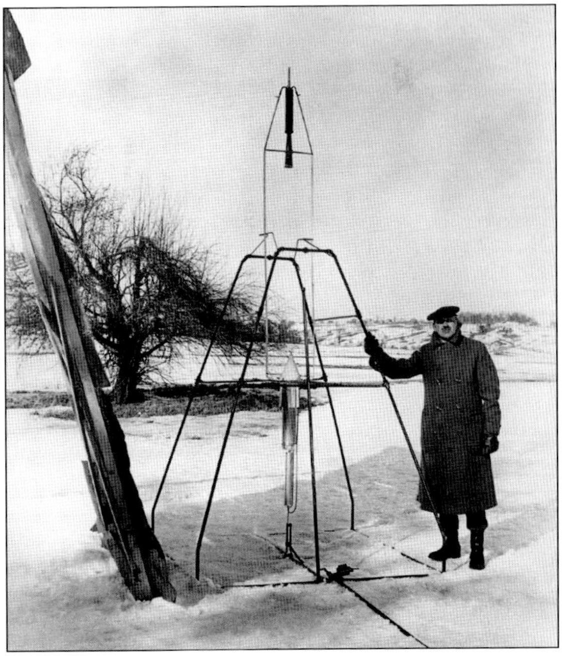

In one of the most iconic photographs in the annals of aerospace history, Robert Goddard stands beside his soon-to-be-legendary rocket. Four people were on hand for Goddard's historic launch of the world's first liquid-fueled rocket on March 16, 1926: Goddard, Esther, Henry Sachs, and Percy Roope.

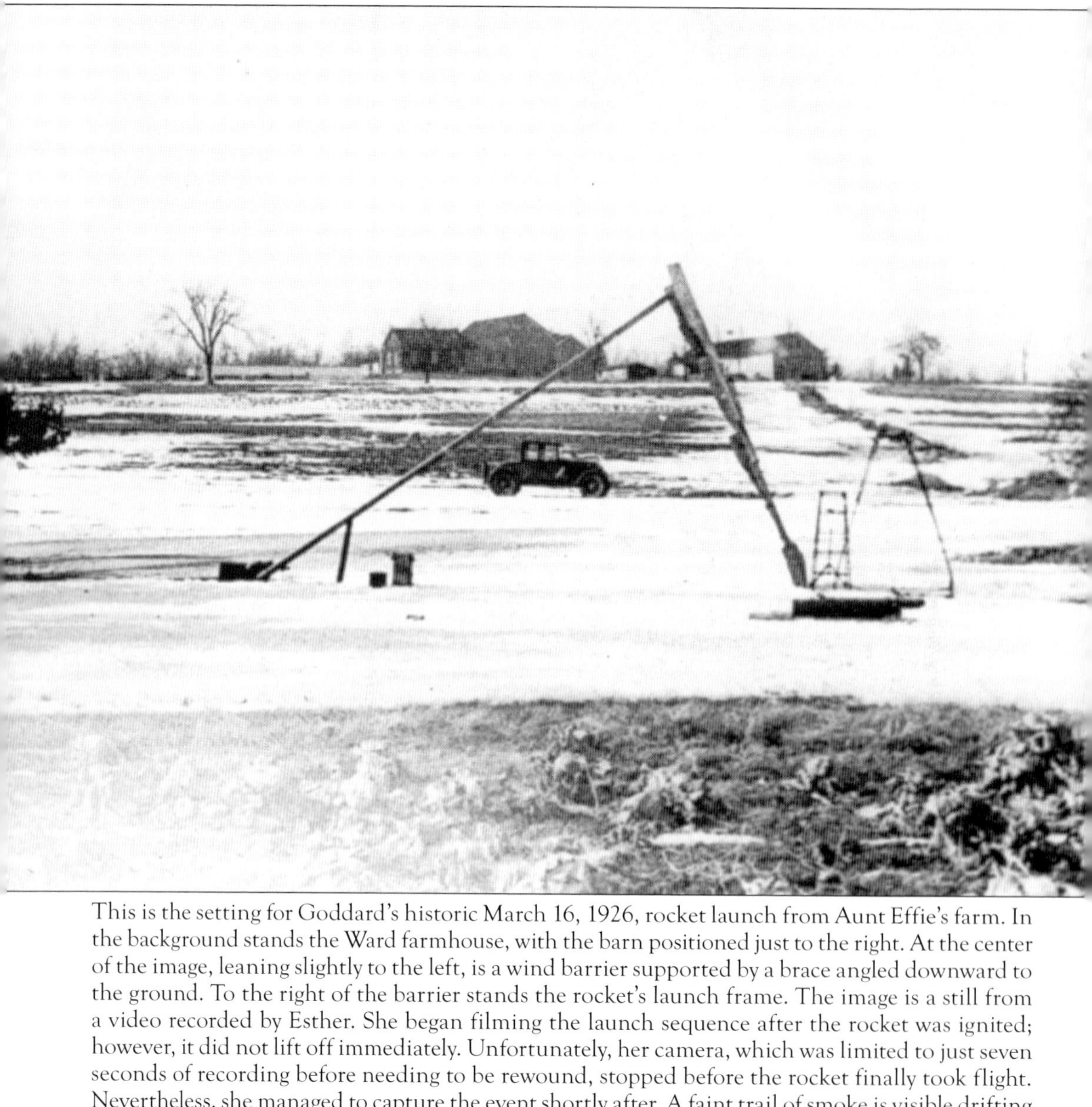

This is the setting for Goddard's historic March 16, 1926, rocket launch from Aunt Effie's farm. In the background stands the Ward farmhouse, with the barn positioned just to the right. At the center of the image, leaning slightly to the left, is a wind barrier supported by a brace angled downward to the ground. To the right of the barrier stands the rocket's launch frame. The image is a still from a video recorded by Esther. She began filming the launch sequence after the rocket was ignited; however, it did not lift off immediately. Unfortunately, her camera, which was limited to just seven seconds of recording before needing to be rewound, stopped before the rocket finally took flight. Nevertheless, she managed to capture the event shortly after. A faint trail of smoke is visible drifting in front of the barn's roof, a subtle testament to the rocket's brief but groundbreaking ascent.

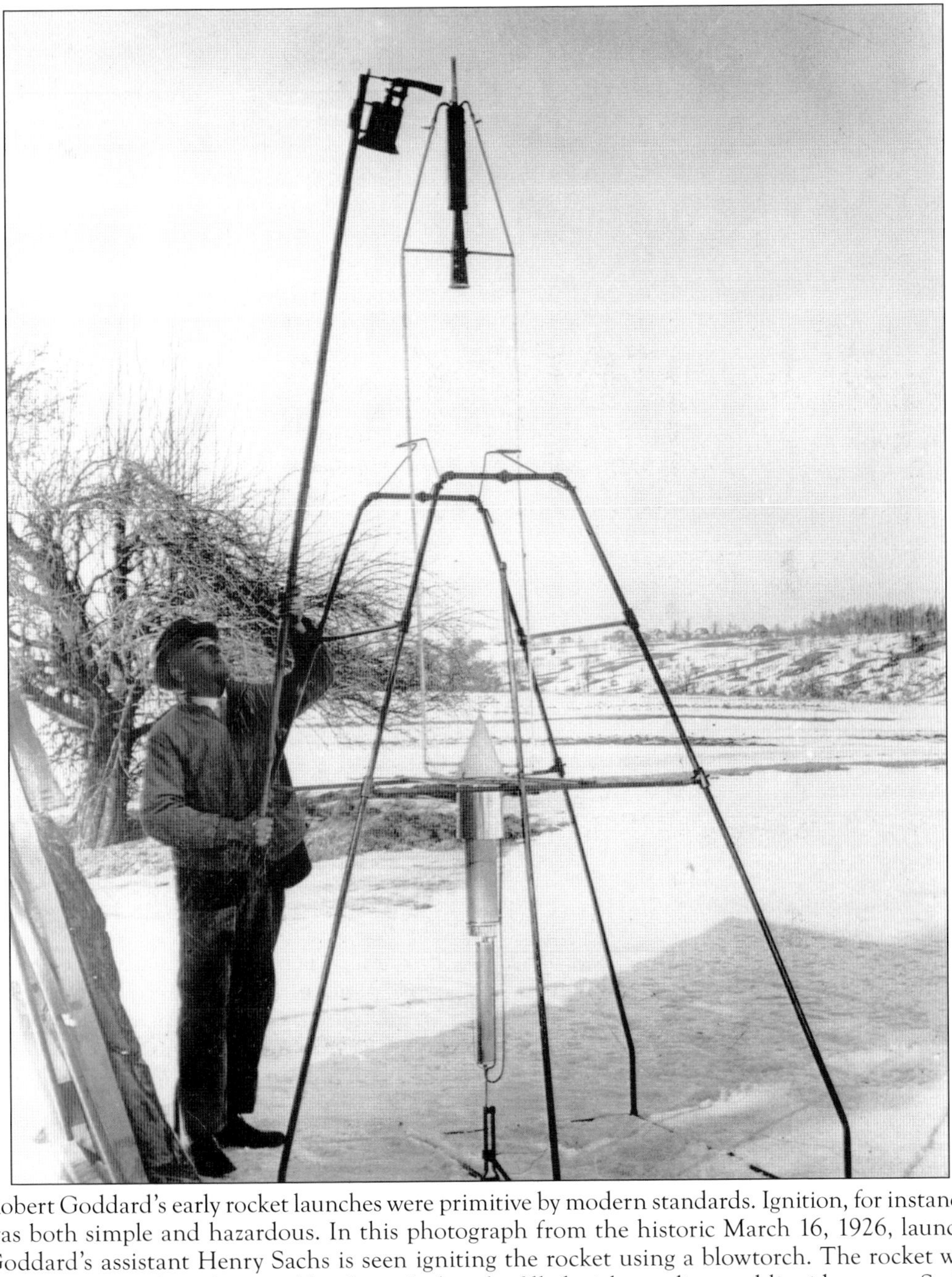

Robert Goddard's early rocket launches were primitive by modern standards. Ignition, for instance, was both simple and hazardous. In this photograph from the historic March 16, 1926, launch, Goddard's assistant Henry Sachs is seen igniting the rocket using a blowtorch. The rocket was mounted atop a launch assembly, above fuel tanks filled with gasoline and liquid oxygen. Sachs had to approach the rocket assembly with a lit blowtorch attached to a long pole, carefully raise the flame to the combustion chamber, turn a valve, and then retreat quickly to safety once the rocket ignited. The rocket did not launch immediately, burning off excess fuel before finally lifting off. It reached a maximum height of 41 feet as it traveled 184 feet downrange during 2.5 seconds. As Goddard noted the next day in his diary, "It looked almost magical as it rose." Safer and more efficient ignition systems, such as wire fuses and remote triggers, would soon replace this blowtorch method.

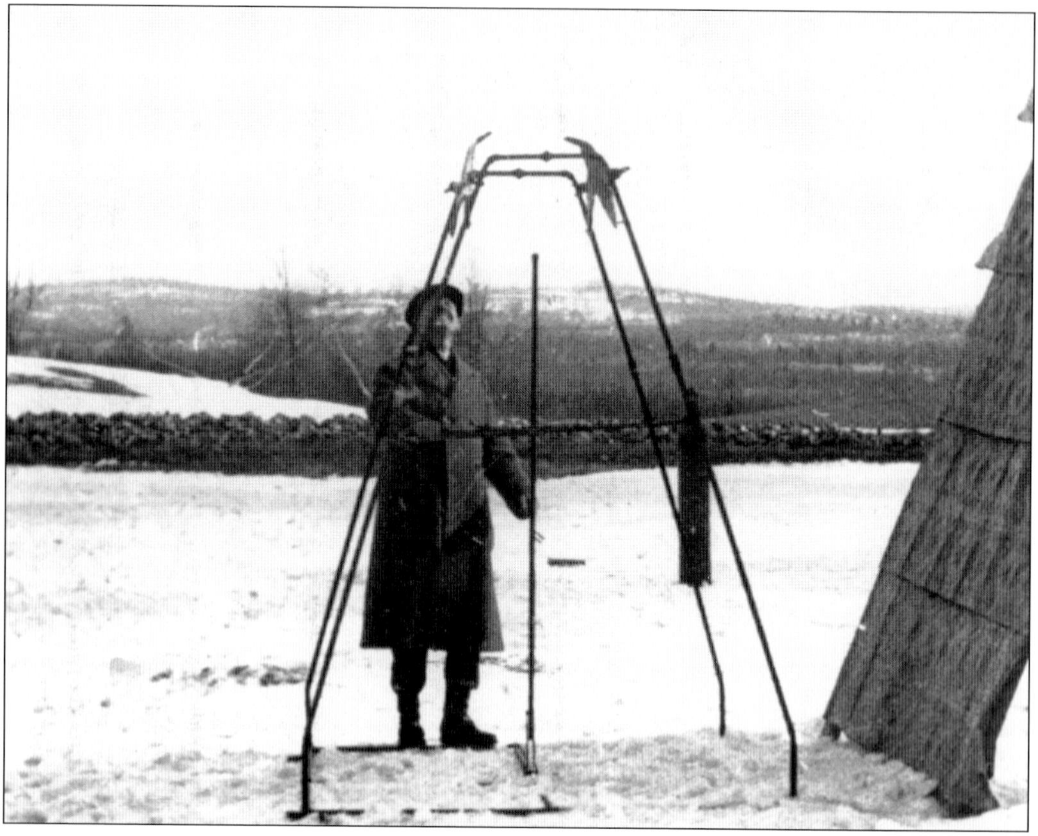

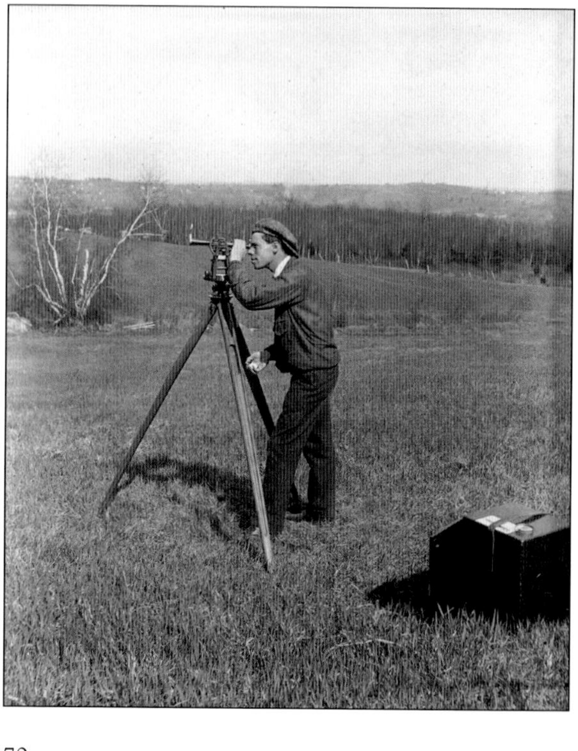

In this photograph, Goddard looks up past the empty launch frame after the rocket finished its short but historic journey. In a report to the Smithsonian Institution's Charles Abbot, Goddard described this picture: "I am enclosing an enlargement made from a film taken at the time of the test of March 16, which I might entitle 'The Empty Frame,' inasmuch as I have been working to make a liquid-propelled rocket leave a frame since 1920."

Percy Roope, an assistant physics professor at Clark University, played a key role in documenting the results of Goddard's early rocket launches. On March 16, 1926, Roope used a theodolite and a stopwatch—standard tools in surveying at the time—to measure the rocket's altitude and trajectory. Roope continued to assist Goddard in subsequent research, helping to validate the performance of Goddard's ever-more advanced rockets.

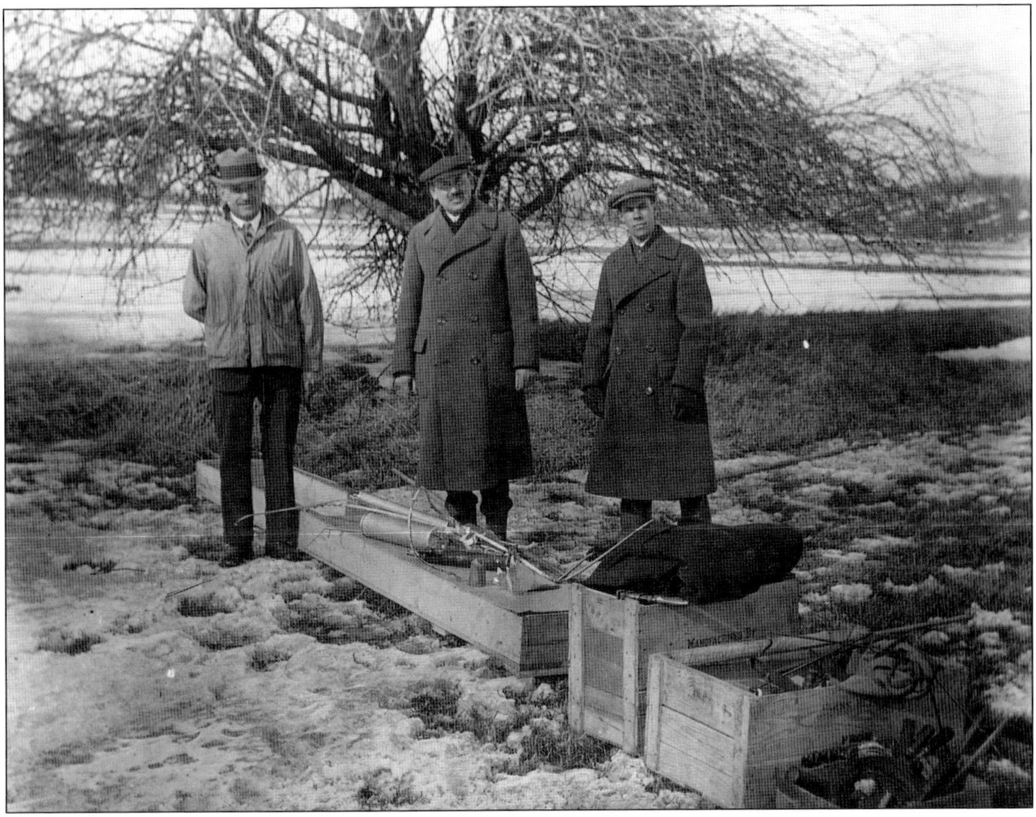

Henry Sachs, Robert Goddard, and Percy Roope gathered the rocket and launch frame parts after the historic March 16, 1926, launch at Aunt Effie's farm. They returned to Clark University, where Goddard analyzed the flight and began refining his designs. Less than a month later, he launched a modified version of the rocket from the same site, continuing his relentless pursuit of more efficient flight.

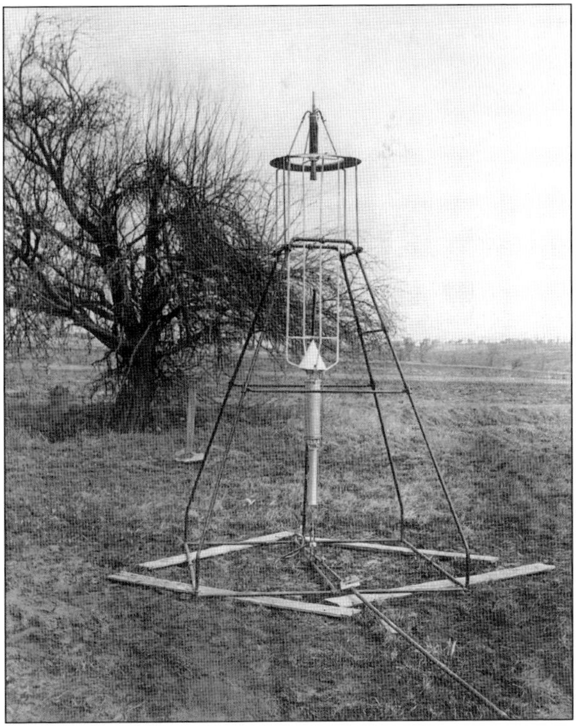

Goddard launched the second liquid-fueled rocket on April 3, 1926. Like any good scientist who tests, refines, and documents, Goddard looked to replicate his first experimental flight. He accomplished this using a similar rocket design but with improvements to the launch frame that included reinforced vertical braces. It flew for about 4.5 seconds and landed 50 feet from the frame.

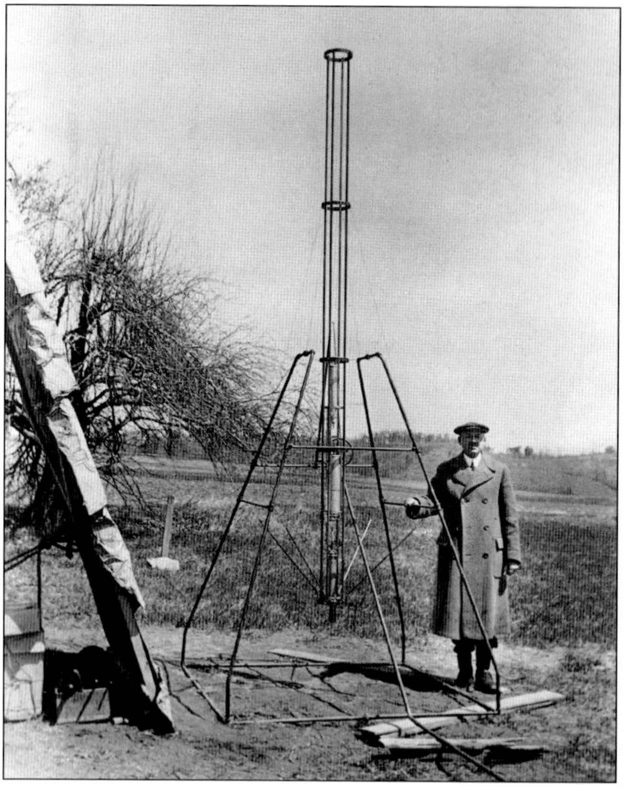

After his second successful launch on April 3, Robert Goddard realized that placing the engine below the tanks, rather than above, did not reduce stability as he had originally believed it would. Such an alignment made for an easier design, easier access for maintenance, better fuel flow based on gravity assistance, and improved propulsion. For all his subsequent designs, such as the one shown here in May 1926 that he did not successfully launch, Goddard used this now-standard configuration.

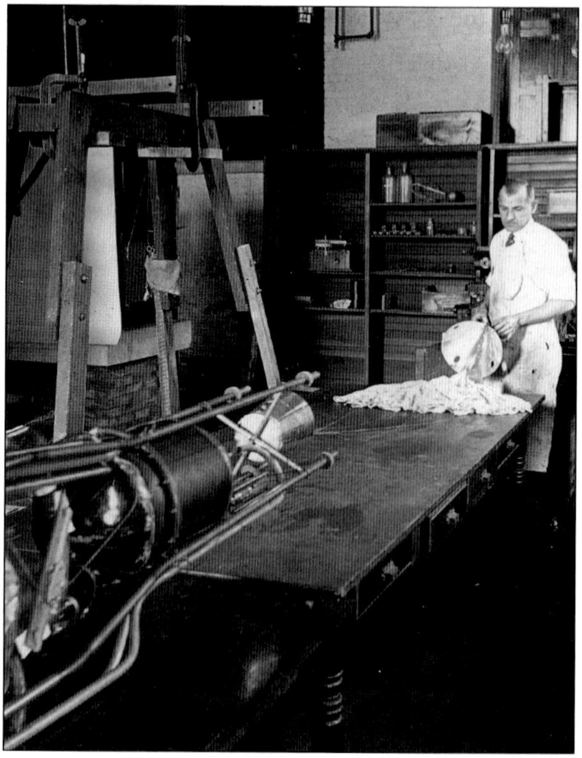

In 1927, Goddard began experimenting with parachutes to ensure that rockets—and later, scientific instruments and other payload items—could return safely to Earth rather than crashing. In this photograph taken in the Clark University physics lab, Goddard's assistant Henry Sachs stretches out a prototype designed to deploy from a rocket during flight.

In addition to parachutes, Robert Goddard devised and tested a variety of other rocket elements as he strove to refine the mechanics of rocket flight, looking for better ways to propel, stabilize, and recover his rockets. He spent most of his time designing and building the components and rockets, such as this 1928 model, in the physics lab, then conducting launches at Clark University and certain other tests at Aunt Effie's farm.

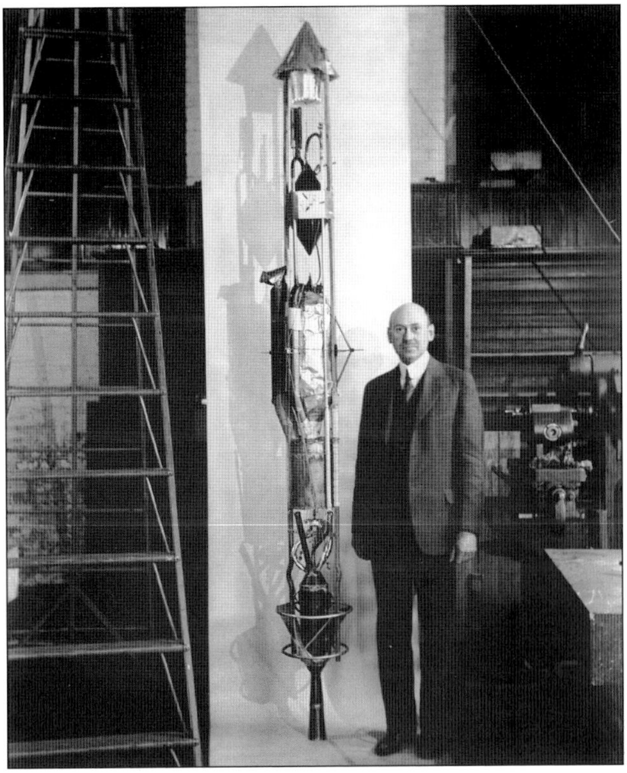

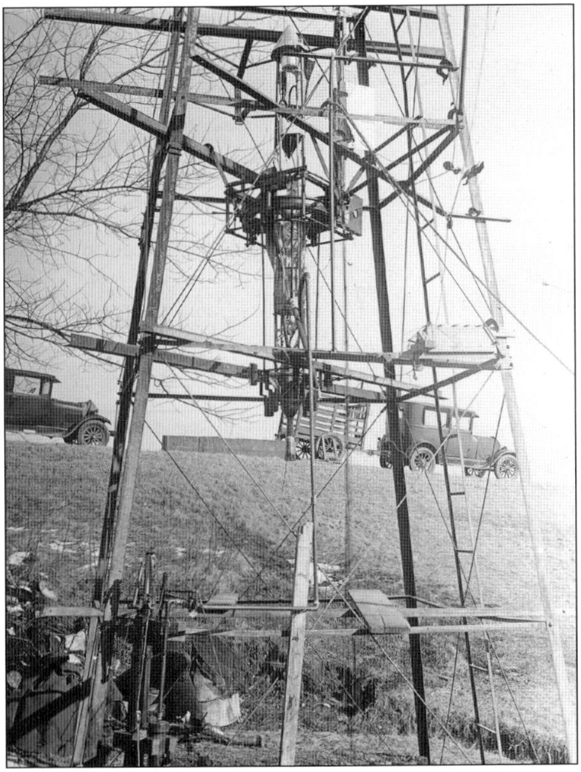

Between 1926 and 1929, Goddard successfully launched four rockets from the farm (several other attempts failed), and he also conducted dozens of other tests, such as the one depicted in this February 6, 1928, photograph, that did not end with an actual launch. In this case, the liquid tank exploded before the rocket could launch.

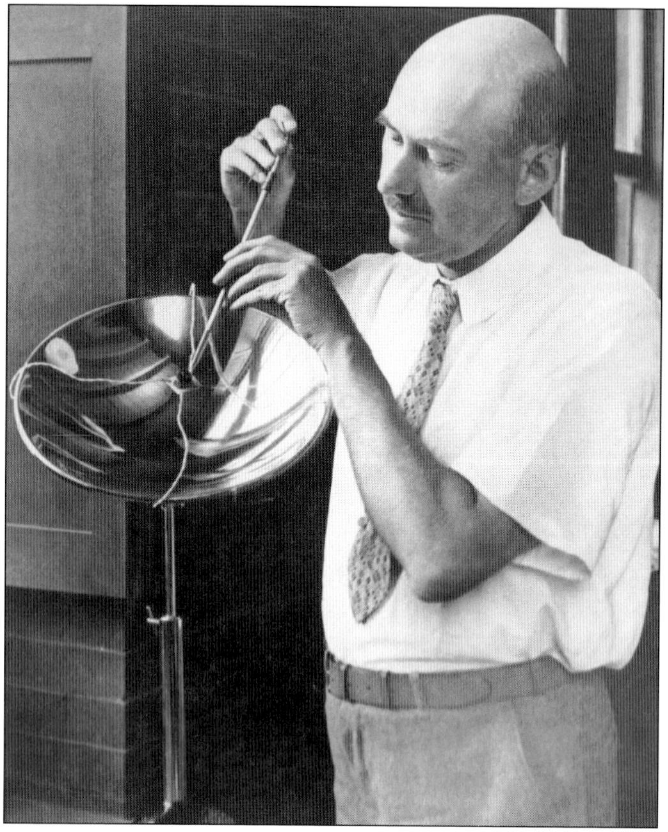

While Robert Goddard is best remembered for his pioneering work in rocketry, he also explored other concepts that were well ahead of their time, such as ion propulsion, interstellar migration, and suspended animation for space travel. He also conducted research in solar energy and its potential as a power source for both space-based and terrestrial applications.

In this photograph, Goddard pulls a trailer loaded with what appears to be parts of his Hoopskirt Rocket, as depicted on the cover of this book. He had tried launching this rocket several times before finally finding success on December 26, 1928. This was his third successful rocket launch.

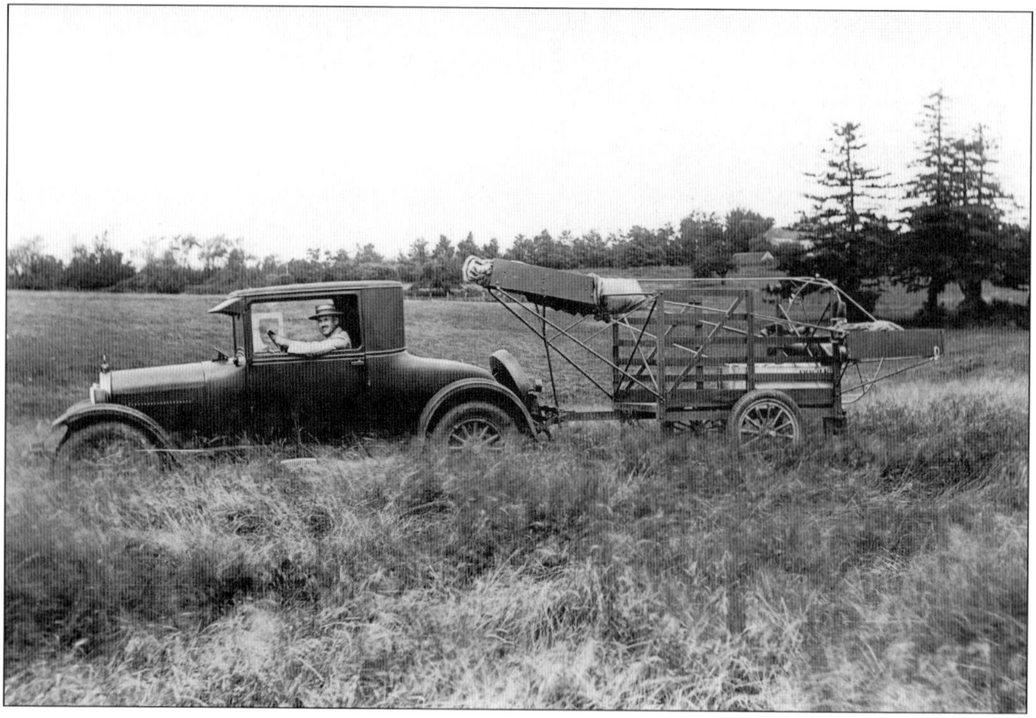

Robert Goddard designed the Hoopskirt Rocket, featuring a steel and aluminum framework not so much for durability or aerodynamics but for testing the liquid-fueled motor. After its successful launch on December 26, 1928, Goddard and Henry Sachs examined the rocket's battered remains.

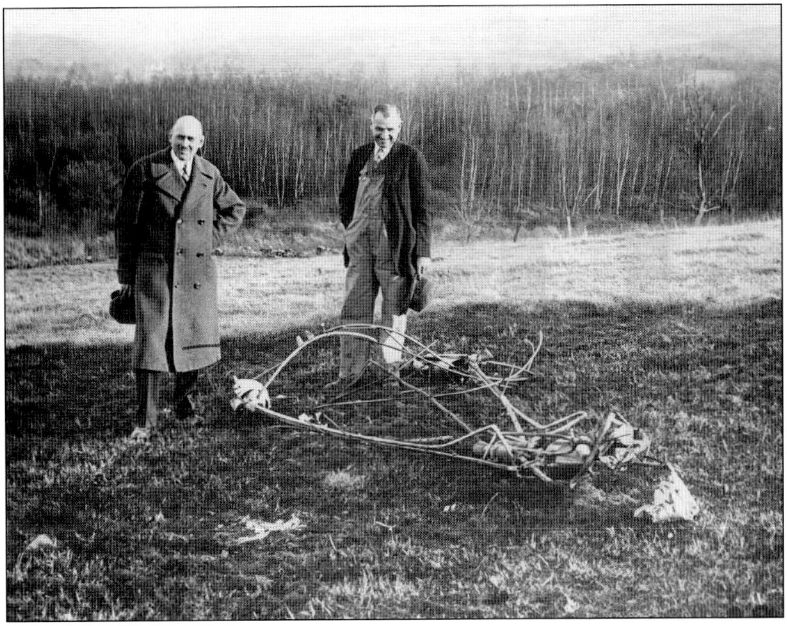

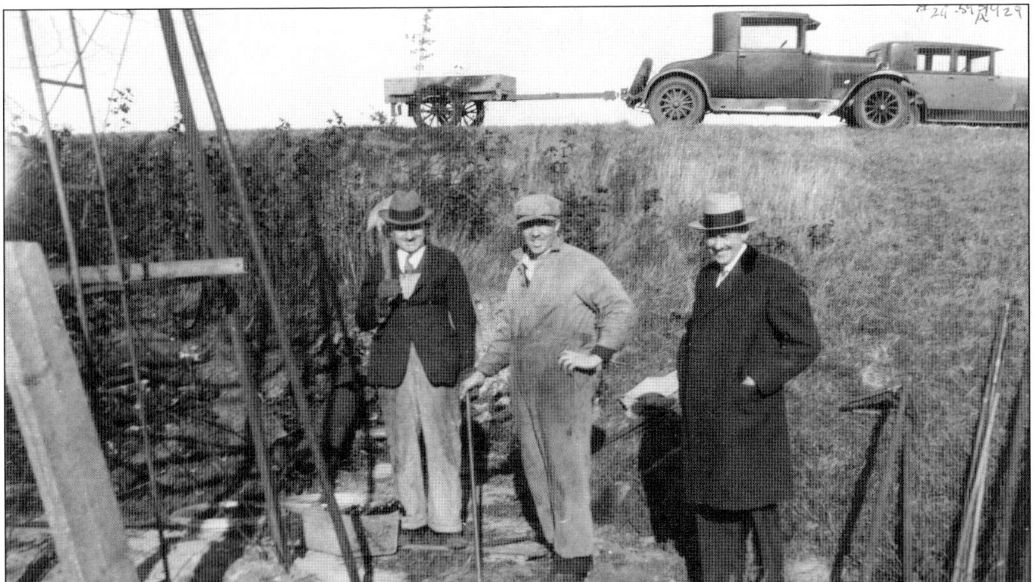

In this 1929 photograph, Goddard (right) stands with members of his team beside the base of a 60-foot launch tower they assembled in order to accommodate larger rockets. It featured a base that could turn in order to change a rocket's orientation and was designed to support rockets with scientific payloads, such as cameras and meteorological instruments.

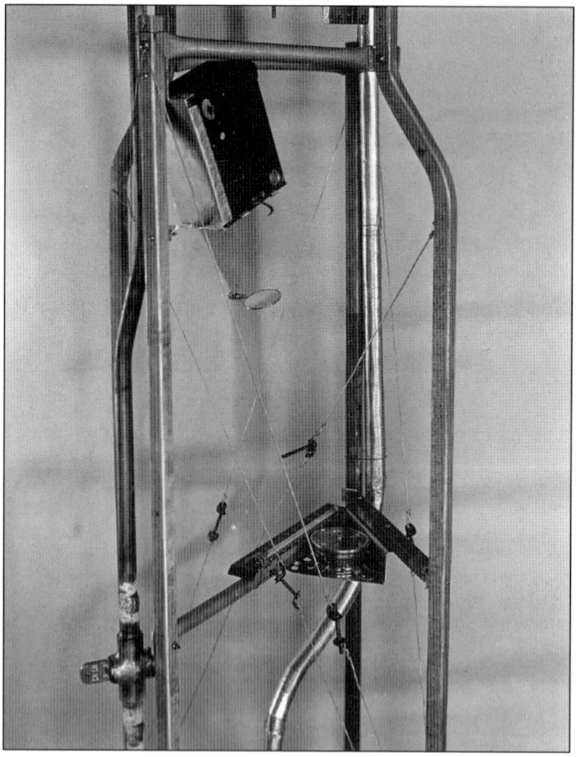

After years of rigorous experimentation and three successful rocket launches, Robert Goddard was ready to take the next step in his rocketry program: adding scientific payloads. In 1929, he attached a camera, thermometer, and barometer to his latest liquid-fueled rocket. He launched the rocket on July 17 at Aunt Effie's farm in Auburn, marking the first time a rocket carried scientific instruments.

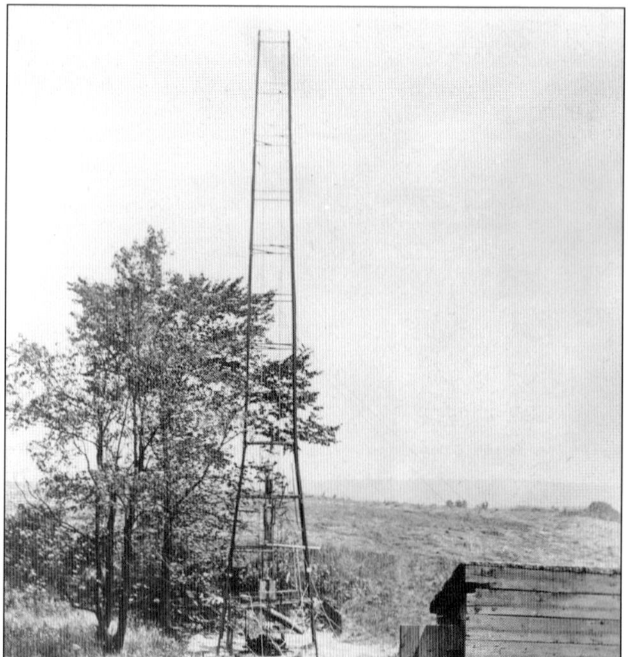

Here is a view of the launch tower and observation shelter on the day of the July 17, 1929, launch. The rocket flew for 18.5 seconds and landed 171 feet from the launch tower. This was the first science payload ever launched on a rocket but also the last of Goddard's liquid-fueled rocket launches from Aunt Effie's farm.

This photograph shows Robert Goddard's team examining the remains of the rocket launched on July 17, 1929, a familiar scene in the early days of rocketry. From left to right are Larry Mansur, Goddard, Henry Sachs, Al Kisk (Goddard's brother-in-law), and Percy Roope. Such landings inspired Goddard to continue experimenting with parachutes on rockets to soften landings and help preserve scientific instruments. This moment captures Goddard's practice of constant experimentation and adjustments based on the results of his tests.

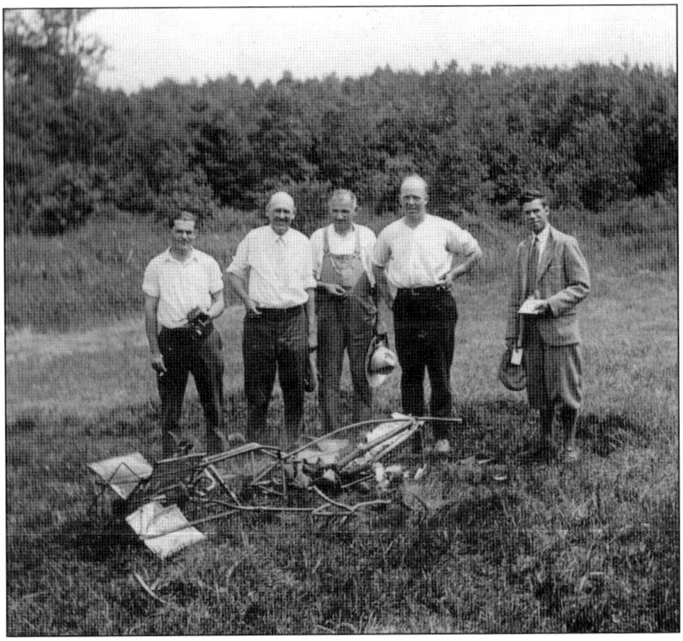

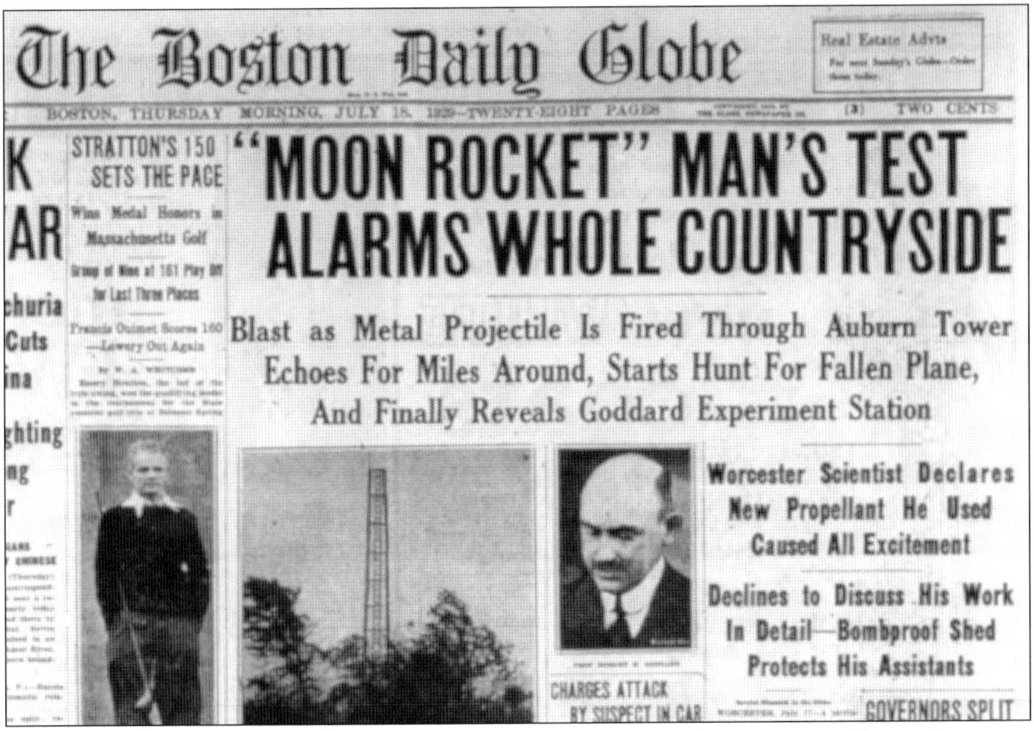

The Boston Daily Globe

Real Estate Advts

BOSTON, THURSDAY MORNING, JULY 18, 1929—TWENTY-EIGHT PAGES

TWO CENTS

STRATTON'S 150 SETS THE PACE

Wins Medal Honors in Massachusetts Golf

Group of Nine at 161 Play Off for Last Three Places

Francis Ouimet Scores 160 —Lowery Out Again

"MOON ROCKET" MAN'S TEST ALARMS WHOLE COUNTRYSIDE

Blast as Metal Projectile Is Fired Through Auburn Tower Echoes For Miles Around, Starts Hunt For Fallen Plane, And Finally Reveals Goddard Experiment Station

Worcester Scientist Declares New Propellant He Used Caused All Excitement

Declines to Discuss His Work In Detail—Bombproof Shed Protects His Assistants

CHARGES ATTACK BY SUSPECT IN CAR

GOVERNORS SPLIT

The thunderous noise of Goddard's July 17, 1929, rocket caught the attention of law enforcement personnel and medics, who rushed to the scene fearing a plane crash. Reporters quickly filed sensational, often mocking stories about his "Moon Rocket." Goddard's assistant Larry Mansur, referencing the popular Edward Sheldon play *Salvation Nell*, quipped, "They ain't done right by our Nell." From then on, Goddard's team nicknamed all their rockets "Nell." (Kevin Schindler personal collection.)

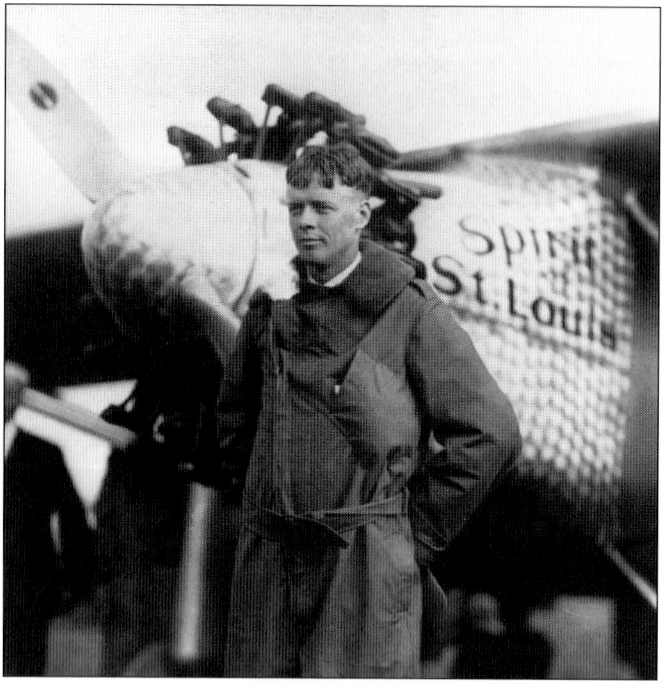

After the July 17 launch, fire officials prohibited Robert Goddard from future launchings at Aunt Effie's farm due to the risk of fire. Goddard soon found a new launch site at the Camp Devens military post. The July 17 launch also captured the attention of famed aviator Charles Lindbergh. He soon befriended Goddard and later helped secure funding from the Guggenheim Foundation to support Goddard's research for years. (Library of Congress.)

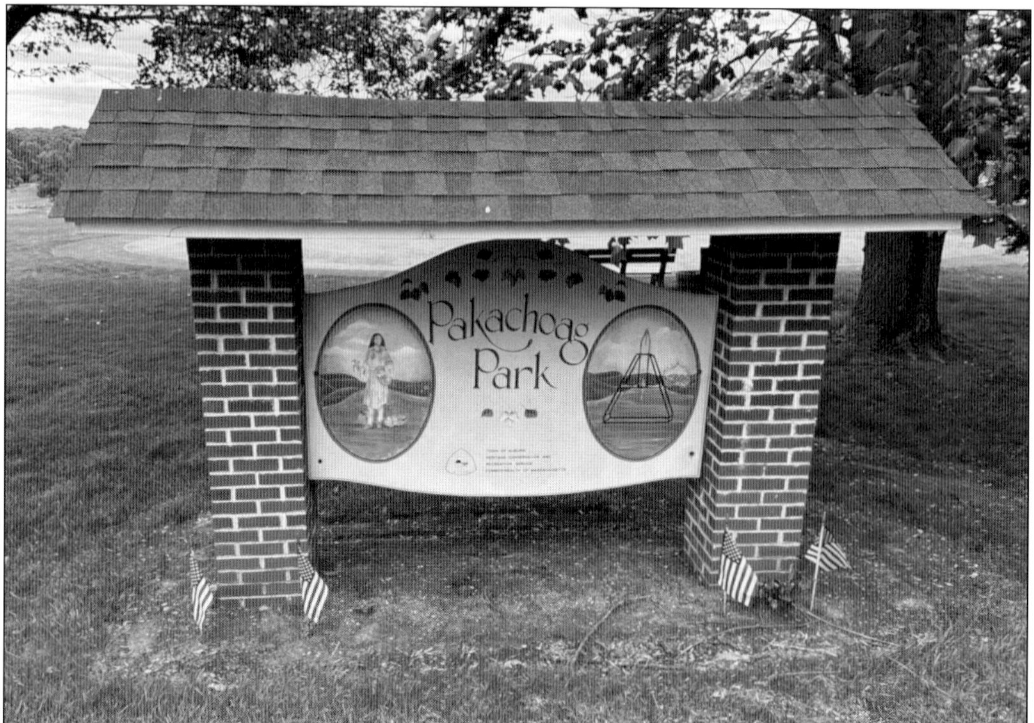

In 1939, a decade after Goddard's final launch at Auburn, Aunt Effie's farm was developed into a nine-hole municipal golf course. Owned by the City of Auburn since 1981, the Pakachoag Golf Course blends the rich historical heritage of the site into its identity as a popular recreational area. (Kevin Schindler personal collection.)

Three historic markers commemorating Robert Goddard and the site's legacy as the birthplace of modern rocketry are nestled within the rolling landscape of Pakachoag Golf Course. The most prominent marker is a granite obelisk that was placed at the approximate location of Goddard's initial launch. Positioned along the fairway of the ninth hole, near a tranquil pond and a small red barn, the obelisk stands as a solemn tribute to a visionary mind. The image at right shows Esther in 1966, wandering around the site that she helped make famous. The image below offers a contemporary view of the monument, now enclosed within a square-shaped garden filled with vibrant flowers—a peaceful blend of nature and history. It serves as an elegant reminder of the humble beginnings of rocketry and space travel. (Right, Robert H. Goddard Library, Clark University; below, Kevin Schindler personal collection.)

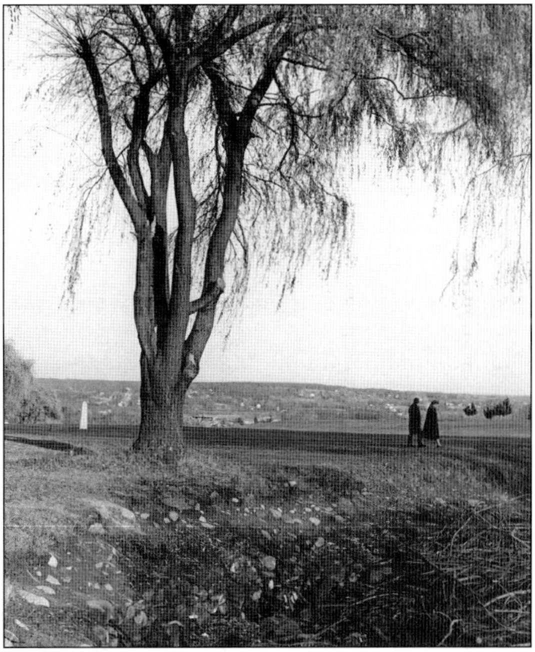

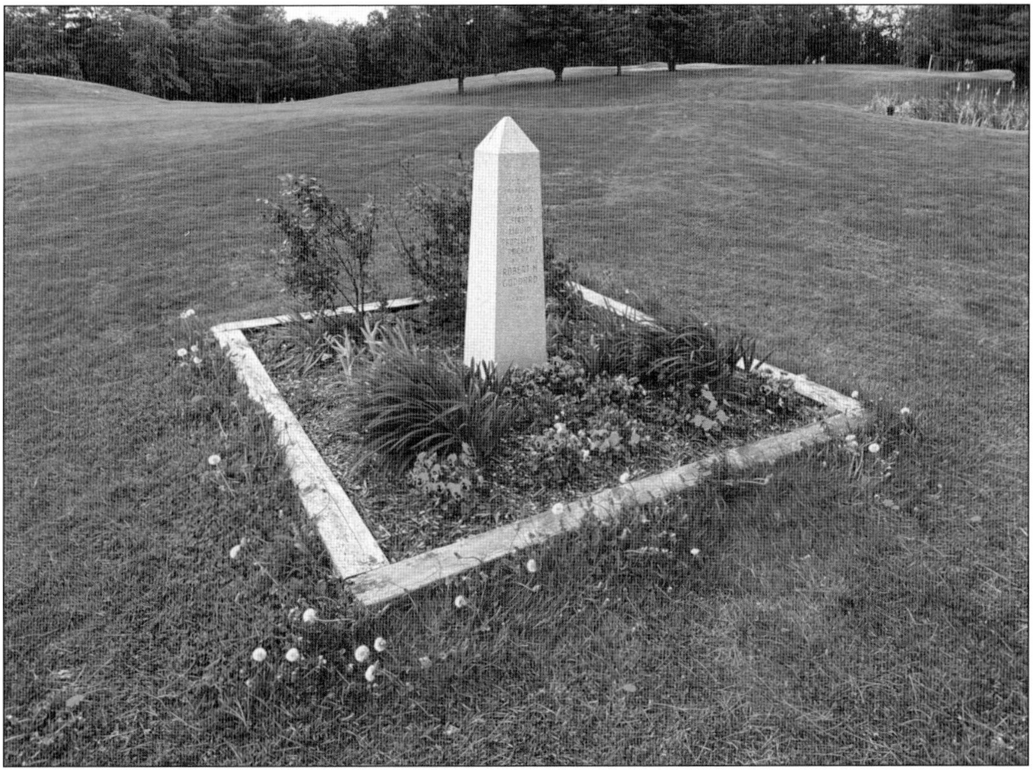

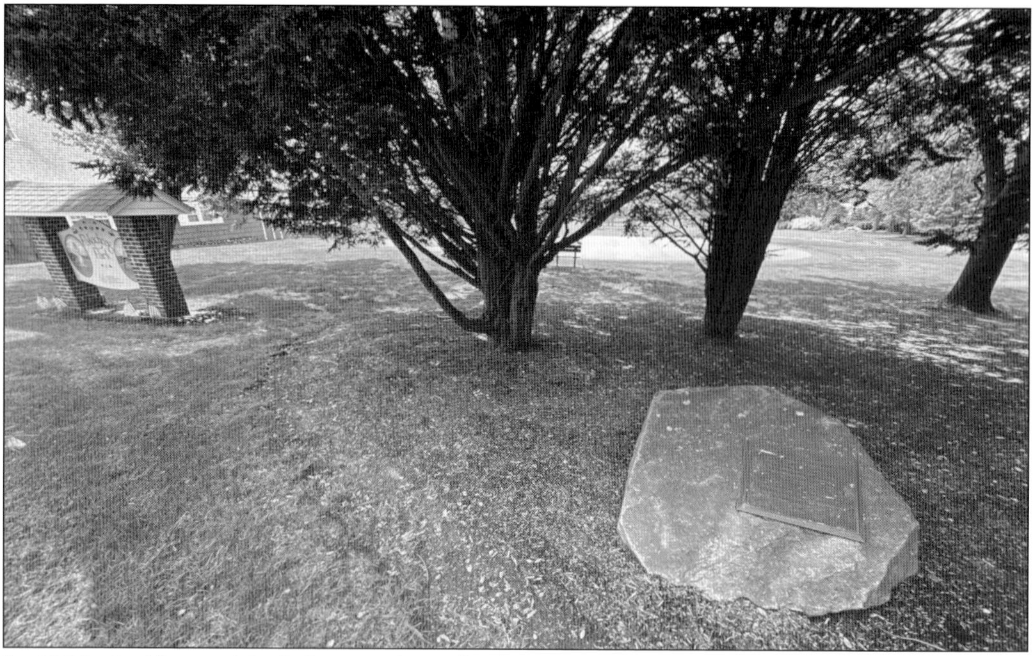

In 1966, the National Park Service formally designated the site of Goddard's groundbreaking rocket experiments as a registered national historic landmark, recognizing its pivotal role in the development of modern rocketry. Officially named the Pakachoag Hill Goddard Rocket Launching Site, this quiet hilltop in Auburn had become the cradle of modern rocketry, where Goddard successfully launched the world's first liquid-fueled rocket on March 16, 1926. A section of the designation application reads, "The launch site itself is located about 1,000 feet southeast of Upland Street (the northwestern boundary of the course), midway between the tee and the green on the ninth fairway. It is marked by a simple granite obelisk approximately four feet high carrying the following inscription: 'Site Of Launching Of World's First Liquid Propellant Rocket By Dr. Robert H. Goddard—16 March 1926.'" The landmark designation not only recognized the site's enormous scientific significance but also ensured its preservation as a place of inspiration and education for future generations. In essence, it is the "Kitty Hawk of the Space Age." (Both, Kevin Schindler personal collection.)

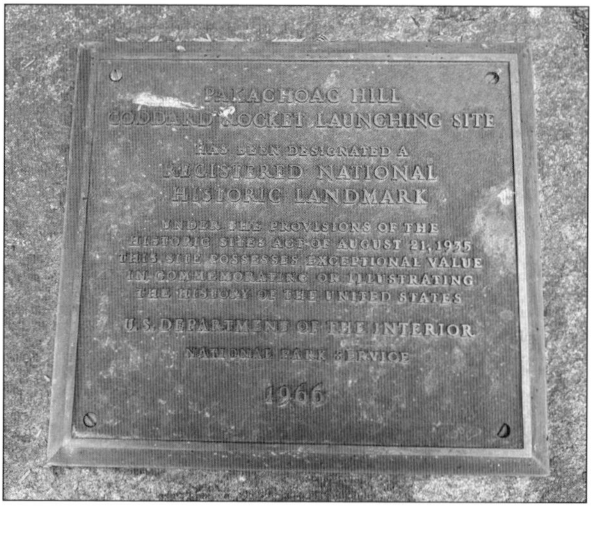

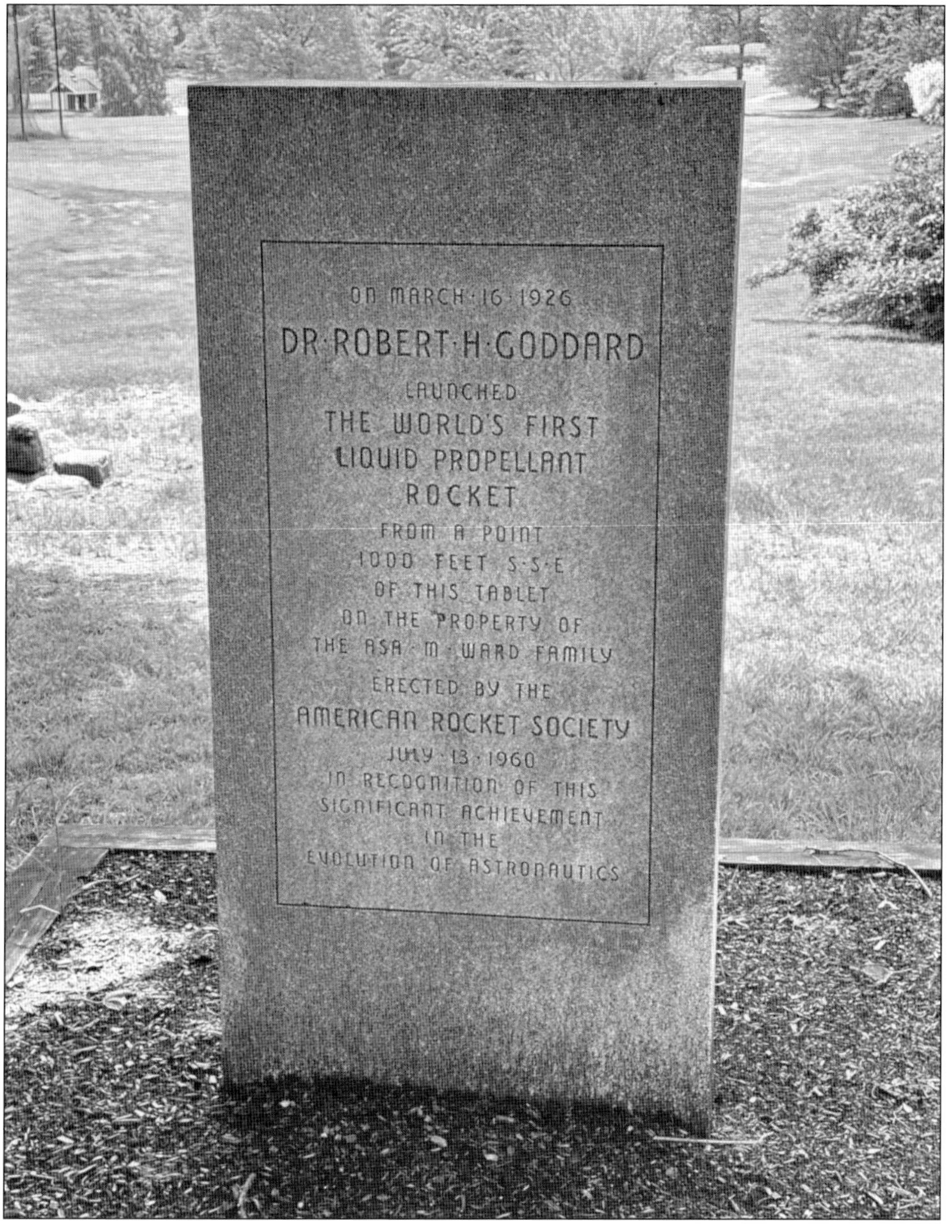

A third historic marker adds depth to the legacy of Robert H. Goddard at Pakachoag Golf Course and stands near the registered national historic landmark plaque that commemorates the site of his pioneering rocket experiments. This particular marker—a stone tablet sponsored by the American Rocket Society—was formally dedicated on July 13, 1960, in recognition of Goddard's groundbreaking contributions to modern rocketry. The American Rocket Society, a professional organization devoted to advancing aerospace science and engineering, chose this site to honor Goddard's role in shaping the future of space exploration. Their installation of this stone tablet reflects the deep respect held by the aerospace community for Goddard's momentous research. Together, all three monuments at Pakachoag Golf Course celebrate the legacy of the man who dreamed of space travel and spent his lifetime laying the steps to achieving that vision, a legacy that continues to inspire generations of scientists and dreamers.

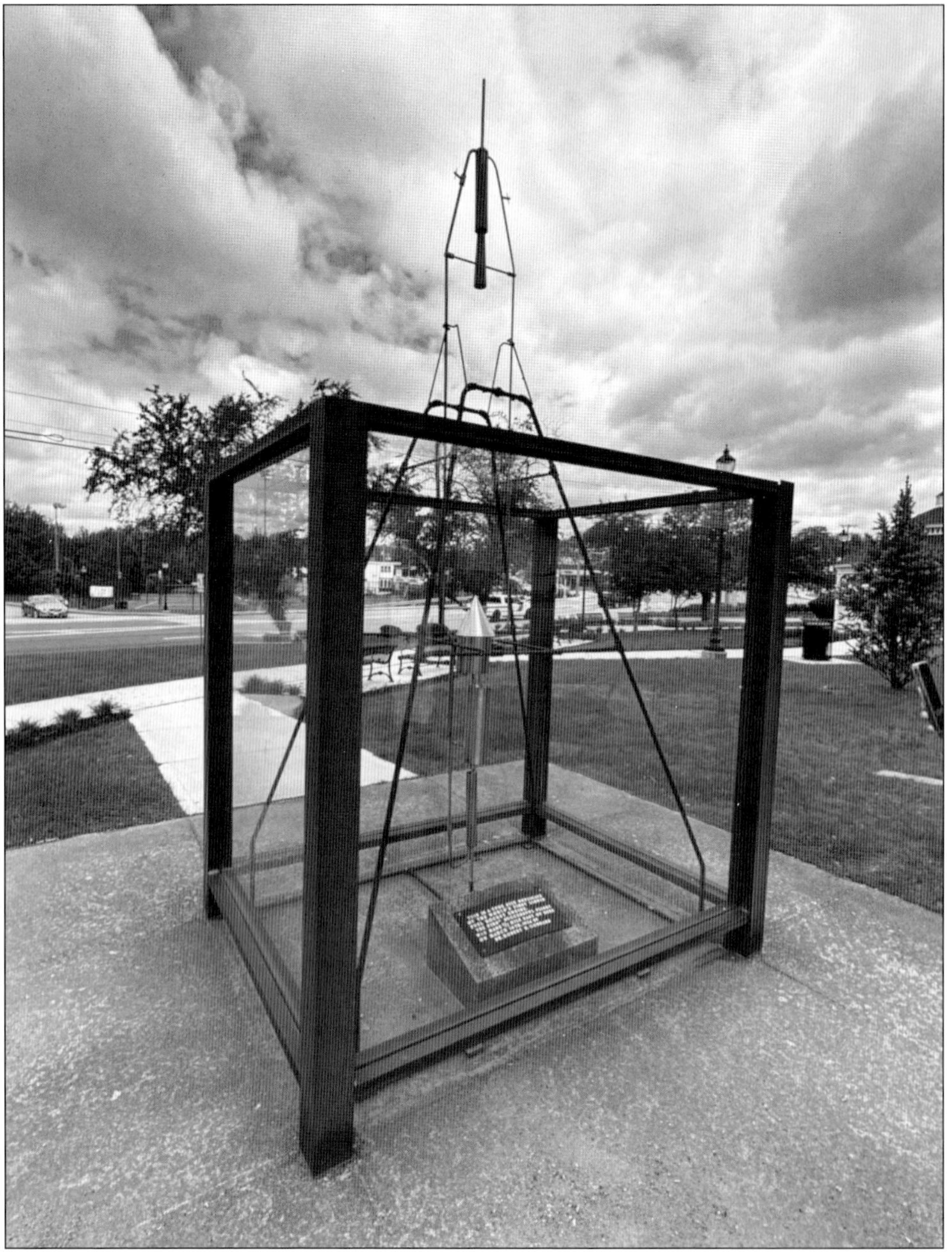

In 1970, the Auburn Rotary Club dedicated Dr. Robert H. Goddard Park, located less than two miles southwest of Pakachoag Golf Course, near the site of Goddard's groundbreaking liquid-fueled rocket launches. This two-acre park serves as a tribute to the father of modern rocketry and features a striking centerpiece: a full-scale replica of Goddard's original 1926 rocket, housed within a protective plexiglass-walled, open-top enclosure. In the mid-2020s, a tall silhouette panel of Goddard holding a rocket replaced a Polaris missile that previously stood on the site. The park also offers a peaceful setting for reflection and learning with benches, a picnic table, a donor recognition plaque, and a commemorative plaque honoring Goddard. Together, these elements create a space that celebrates both the scientific legacy and the visionary spirit of a man whose innovations helped propel humanity into the space age.

Four

Camp Devens
Brief Interlude

Sic itur ad astra ("Thus one goes to the stars")

—Inscription on monument at Camp Devens
recognizing Robert Goddard's on-site rocketry research

Seeking a new location for his experiments after leaving Aunt Effie's farm in Auburn in July 1929 and desiring greater privacy, Robert Goddard turned to a US Army facility located 28 miles northeast of his home in Worcester. Known then as Camp Devens—later renamed Fort Devens—the site appeared remote and spacious enough to accommodate his work.

With assistance from Charles Abbot of the Smithsonian Institution, Goddard secured permission to use a section of Camp Devens near a water hole ominously named Hell's Pond. His team transported the 60-foot launch tower from Auburn to the new site and repurposed an old chicken coop as their makeshift workspace.

Goddard quickly discovered that Camp Devens was far from ideal. Testing was only permitted after rainfall or snowfall (when the ground was sufficiently damp), the rough access road made it nearly impossible to move equipment without risking damage, and the ongoing training of soldiers nearby hampered his efforts. Although he managed to conduct several experiments and static tests—particularly focusing on rocket chambers—beginning in late 1929, Goddard never launched any rockets from the site.

By the spring of 1930, Goddard was grappling not only with the limitations of Camp Devens but also with dwindling financial support from the Smithsonian Institution. Just as his prospects seemed bleak and his future looked dark, a light in the form of Charles Lindbergh was about to illuminate Goddard's path forward.

Fascinated by Goddard's 1929 rocket launch and the broader potential of space flight, Lindbergh paid a visit to Worcester in November of that year, meeting with Goddard and learning about his rocket research. Over the course of the next half year, the enthusiastic Lindbergh shared news of the visit with his friend Daniel Guggenheim and convinced him to fund Goddard's research, initially at a level of $100,000 over four years. Daniel's son Harry administered the fund and would continue to support Goddard for years.

This generous support enabled Goddard to continue his work and seek a more suitable location for his experiments. He chose Roswell, New Mexico, and in 1930 relocated there with his wife, Esther, and his team. Although they returned to Worcester from 1932 to 1934, the move to Roswell effectively marked the end of Goddard's research activities in Massachusetts.

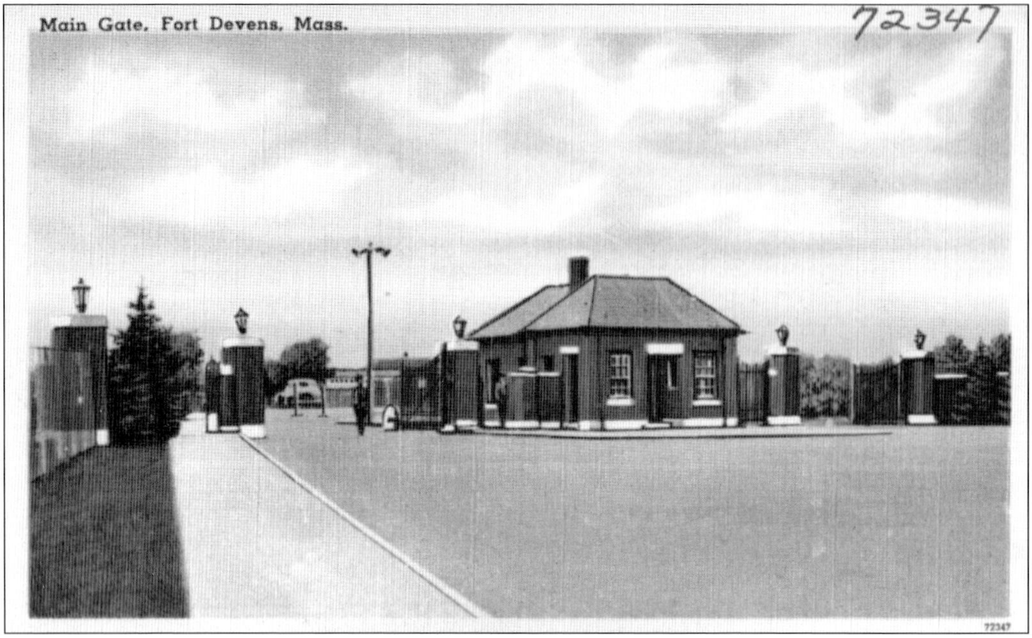

The US Army established Camp Devens in 1917 as a temporary facility for training soldiers during World War I. It was named in honor of Maj. Gen. Charles Devens Jr., a veteran of the American Civil War, a Massachusetts state senator, a Massachusetts Supreme Court justice, and the US attorney general. It was redesignated as a permanent facility in 1931 and renamed Fort Devens the following year. (Kevin Schindler personal collection.)

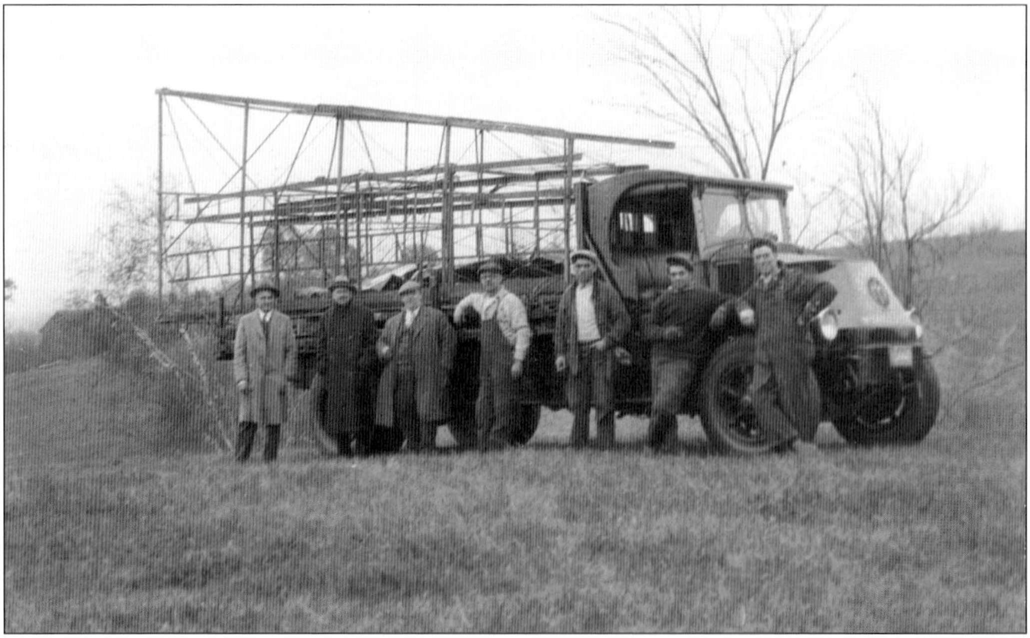

In late October, the War Department granted a six-month license to erect a tower and conduct "flying rocket tests" at Camp Devens. The license was granted to the Smithsonian, but Robert Goddard would conduct the tests. On October 30, Goddard and a team of helpers disassembled the 60-foot tower at Aunt Effie's farm, and on the next day, they drove it to Camp Devens for reassembly.

In early November 1929, Goddard's team reassembled their 60-foot rocket launch tower and converted an old chicken coop into a makeshift operations shack. The site, remote and largely barren except for an abandoned house nearby, offered the solitude Goddard preferred for his experiments. The facility was soon ready for Goddard to resume his testing, but he soon got distracted by a very special visitor that would change the course of his career: Charles Lindbergh. Intrigued by reports of Goddard's July 17 rocket launch, Lindbergh traveled to Worcester on November 23 to meet with Goddard. On the day of the visit, Goddard gave Lindbergh a tour of his lab at Clark University, detailed his research, shared pictures and videos of his launches, and discussed the future of rocketry. The afternoon concluded at the Goddard home, where Robert, Esther, and Charles enjoyed a discussion over chocolate cake and coffee. Impressed by Goddard's work, Lindbergh later persuaded the Guggenheim family, major aviation philanthropists, to fund Goddard's research.

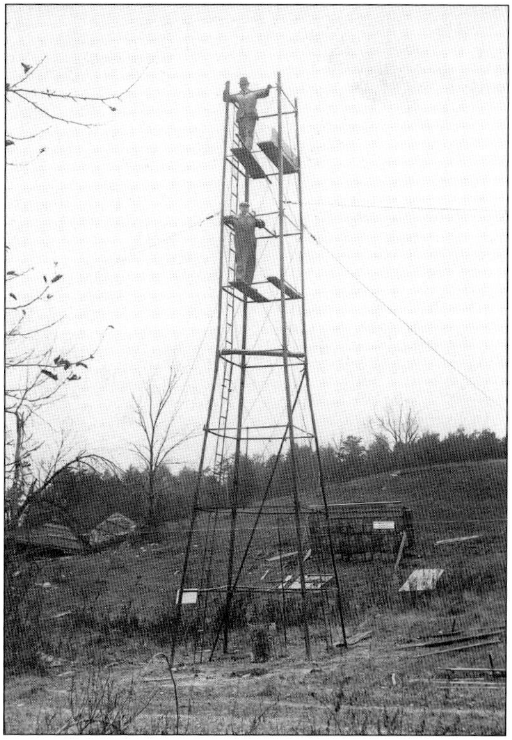

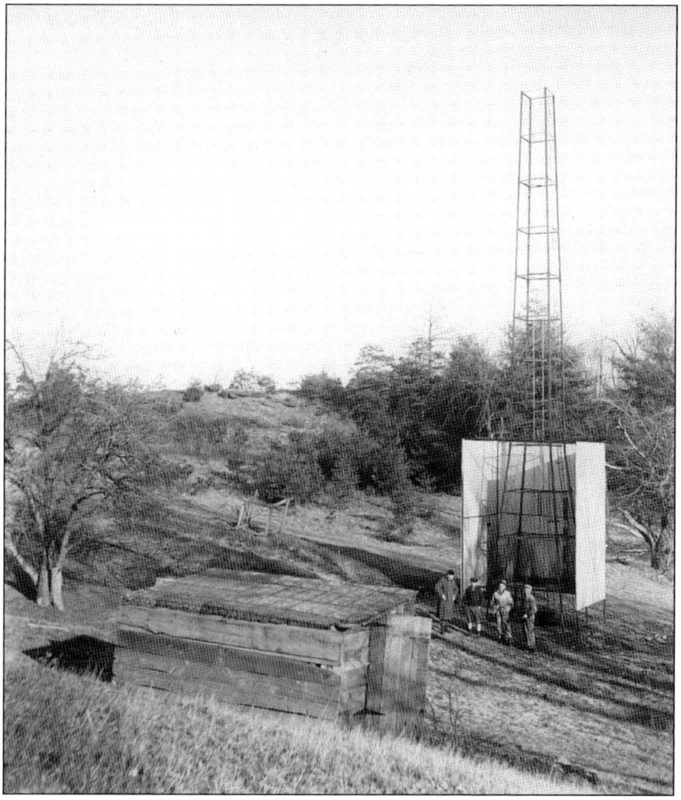

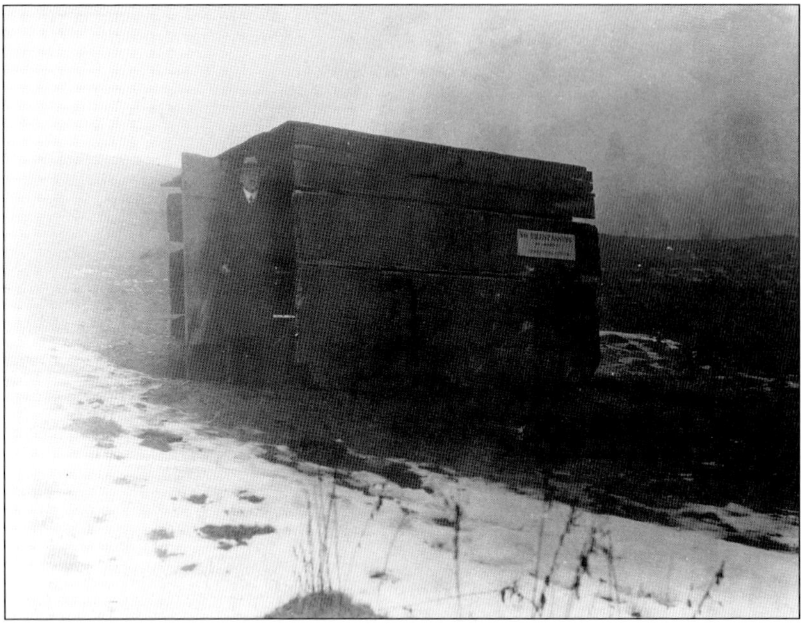

Robert Goddard stands by the "shack" that served as both a preparation area and a shelter for his team during rocket experiments. As part of the War Department's license granting permission to use the site for rocket research, Goddard was required to post "No Trespassing" signs around the perimeter and at key access points.

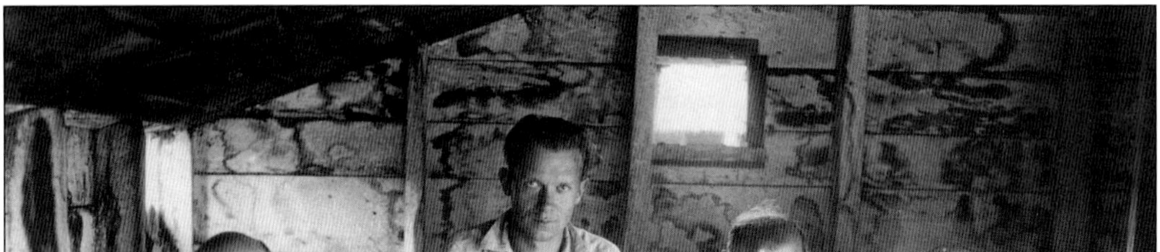

This photograph, taken in the spring of 1930, shows Goddard with his team at the time, from left to right: Goddard, holding his signature cigar, Charles Mansur, Henry Sachs, and Albert Kisk. Though modest in appearance, the shack served as the nerve center for Goddard until funding from the Guggenheim Foundation allowed him to move his testing operation to Roswell, New Mexico, in June 1930.

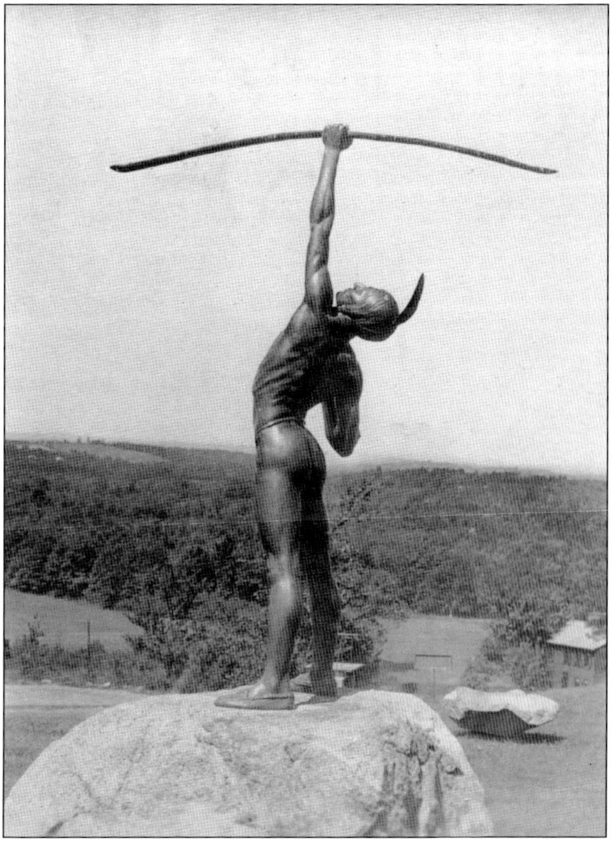

Located just seven miles from today's Fort Devens, the Fruitlands Museum in Harvard, Massachusetts, features a striking statue titled *Pumunangwet*, an Algonquin name meaning "He Who Shoots the Stars." It honors the Native American heritage and symbolizes aspiration and cosmic wonder. Though not directly connected to Robert Goddard, the statue is a reminder of broader humanity's interest in the stars and their inspiration to people around the globe. (Kevin Schindler personal collection).

Although Goddard's rocket testing at Camp Devens was brief, the site proved pivotal to his pioneering research. Three decades after he concluded his experiments there, supporters of his legacy proposed a commemorative monument. On April 2, 1963, Esther joined military officials in a ceremony to dedicate the location of a memorial tower honoring Goddard's contributions to rocketry.

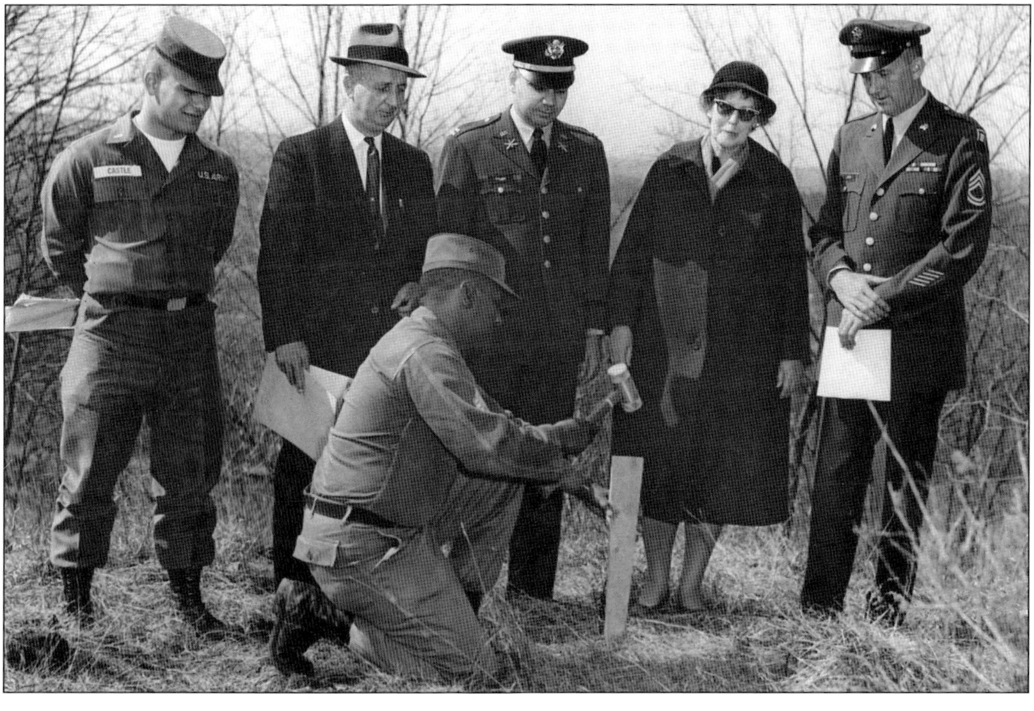

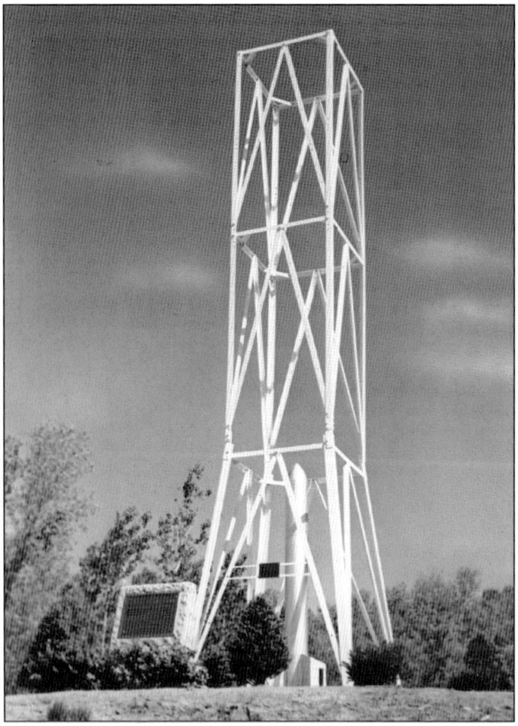

The Dr. Robert H. Goddard Monument at Fort Devens consisted of a tower stylized after the one he used for his experiments at the site. Positioned in front of the tower was a commemorative plaque that detailed the significance of Goddard's research and the historical relevance of the location. At the bottom of the plaque was the Latin phrase *Sic itur ad astra*, meaning "Thus one goes to the stars." This quote originates from the *Aeneid*, the epic poem written by the Roman poet Publius Vergilius Maro, better known as Virgil, between 29 and 19 BC. The phrase has served as an inspirational message of ambition through the centuries, embodying the notion that the path to greatness (reaching the stars) is through vision, perseverance, and dedication, ideals that defined Goddard. Sparked by his cherry tree vision at the age of 17, Goddard spent his life doggedly pursuing his dream of developing the means for spaceflight. The monument, like Goddard himself, served as a symbol of aspiration and the achievement of one's goals.

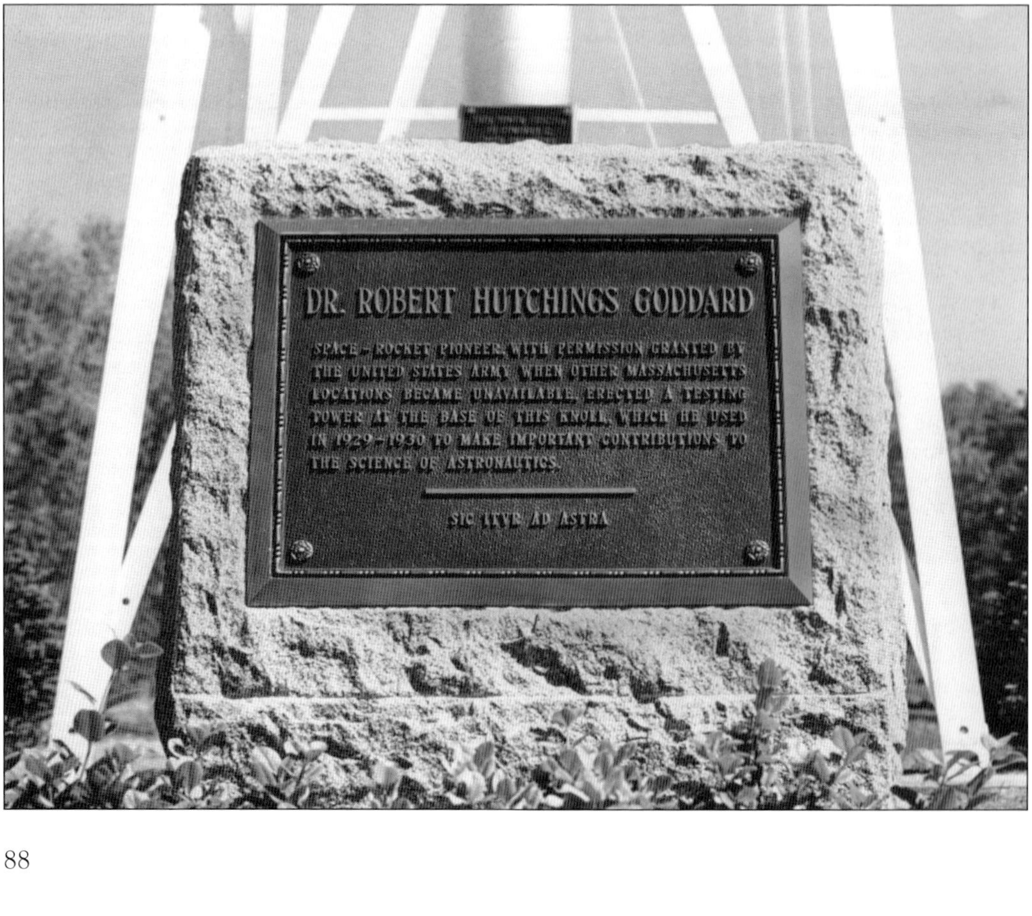

Five

Beyond Massachusetts
New Mexico and Other Stops

Living in our dreams of yesterday, we find ourselves still dreaming of impossible future conquests, as Robert Goddard dreamed, as men have dreamed throughout the ages.

—Charles Lindbergh, preface, *This High Man*, 1963

While this book focuses on Robert Goddard's deep ties to Massachusetts, it's essential to include this bonus chapter highlighting his work beyond the Commonwealth to fully appreciate the scope of his contributions to science and rocketry.

Born in Worcester in 1882, Goddard remained in Massachusetts throughout his college years. In 1912, he moved to New Jersey after accepting a research fellowship at Princeton University. He returned to Worcester in 1913 upon contracting tuberculosis. After recovering, he stayed in Worcester and joined the Clark University physics department, eventually becoming its director. In 1918, he briefly conducted military research at Mount Wilson Observatory in California, developing early solid-fueled rockets for the US Army.

In 1930, with funding from the Guggenheim Foundation secured through Charles Lindbergh's advocacy, Goddard moved his operations to Roswell, New Mexico, where he conducted groundbreaking rocket tests.

Robert and his wife, Esther, lived in Roswell until 1932, when funding from the Guggenheim Foundation was cut, a victim of the Great Depression. The Goddards moved back to Worcester until 1934, when Guggenheim funding was reinstated and the Goddards returned to Roswell. They stayed there until 1942, when they moved to Maryland upon Robert accepting a research position with the US Navy working on JATO (jet-assisted take-off) technology.

Robert died in 1945, and Esther returned to Worcester, dedicating the rest of her life to preserving his legacy.

What follows here are highlights of Robert Goddard's work beyond the borders of Massachusetts, especially his important research in Roswell.

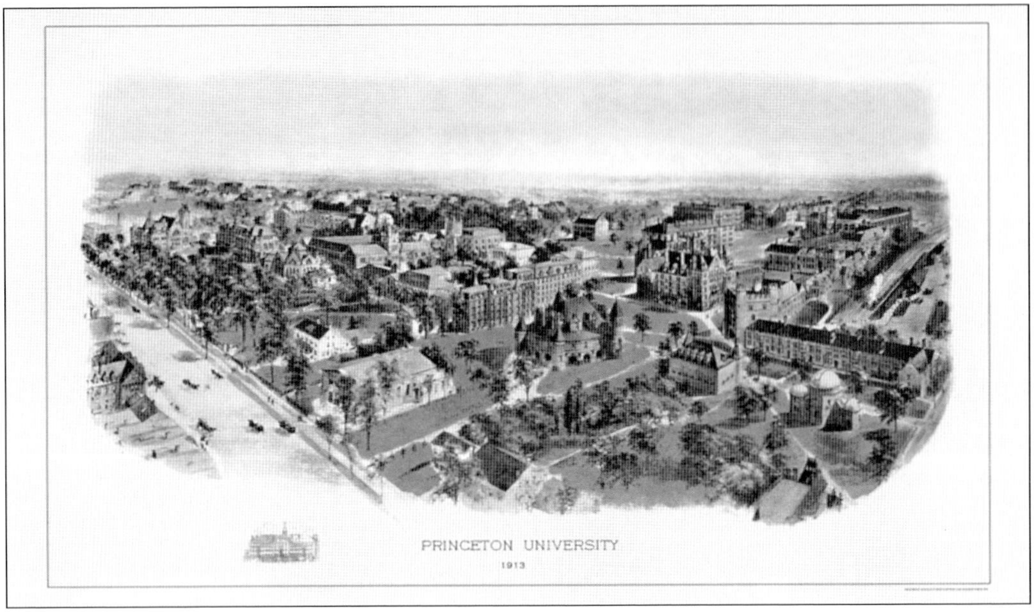

In 1912, Goddard moved to New Jersey to work as a research fellow at Princeton University's Palmer Physical Laboratory. He left this position the following year when he became ill with tuberculosis. He moved back home to Worcester to recover, and in 1914, he accepted a position at his alma mater, Clark University, where for most of the next two decades he taught classes and carried out research in rocketry. (Kevin Schindler personal collection.)

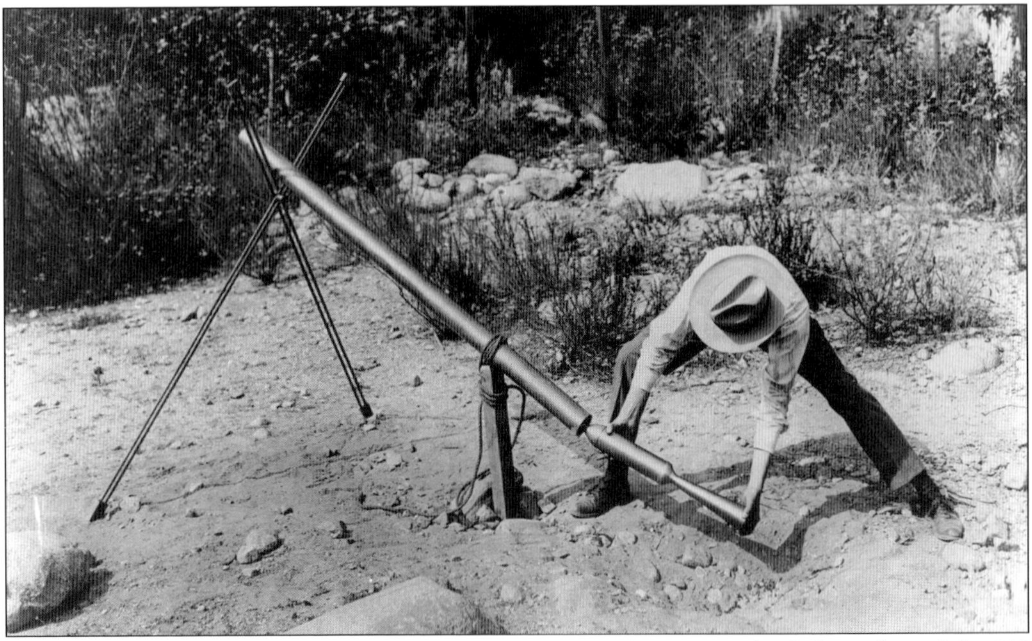

With the onset of World War I, Robert Goddard reached out to multiple branches of the US military, offering insights into how his pioneering rocketry research could be adapted for defense purposes. In 1918, while working part-time at Mount Wilson Observatory in California, he applied his expertise in rocket propulsion to develop a tube-based launcher, which would later evolve into the bazooka.

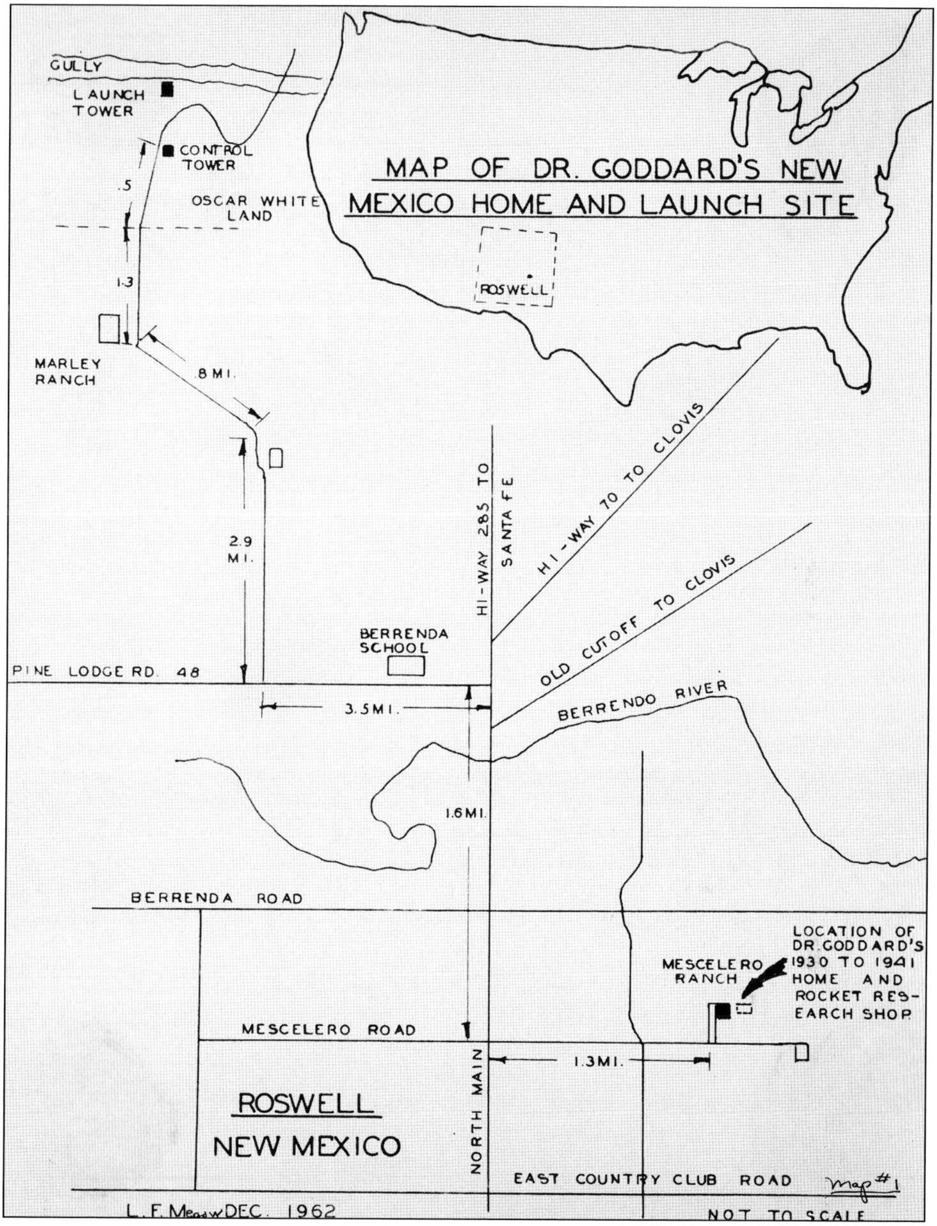

For many years, Goddard received modest financial support from institutions such as the Smithsonian Institution to pursue his groundbreaking research in rocketry. However, in 1930, his work gained a significant boost when the Guggenheim Foundation began funding his efforts. This new backing allowed Goddard to expand his research efforts and seek a more suitable location for rocket launches. To determine the ideal site, he consulted with Lindbergh and meteorology professor Charles Brooks. Lindbergh recommended the American Southwest, while Brooks advised selecting a location with flat terrain, a stable climate, and reliable transportation access. Taking their advice into account, Goddard chose Roswell, New Mexico, as his new base of operations. In July 1930, he and Esther relocated to the area, settling at a place called Mescalero Ranch. This would serve both as their home and as a research center for Robert's work. About a dozen miles northwest of the ranch, Robert established a launch site where he could conduct his experiments in relative isolation.

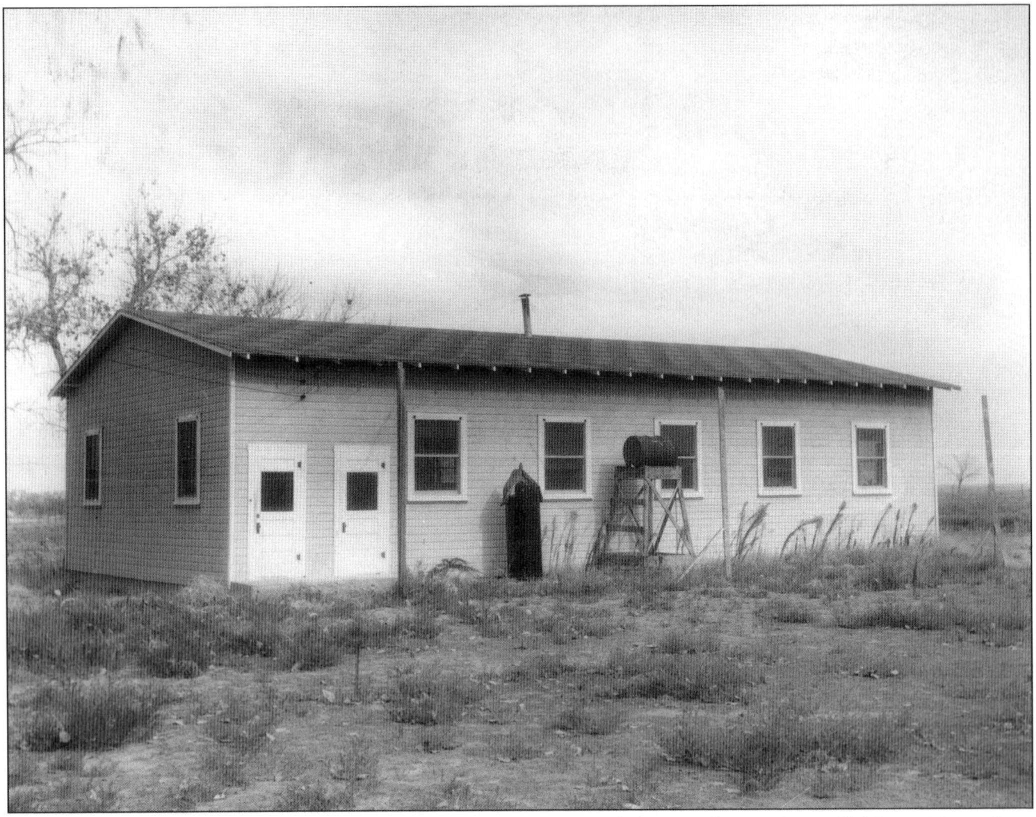

Soon after arriving in Roswell, the Goddards found a ranch house, located at 1501 East Mescalero Road, that they could rent. Robert Goddard and his team spent much of August 1930 designing and building several structures near the ranch house. They first erected a machine shop, shown here, followed by a tower to be used for static tests and a shelter to use while conducting the tests.

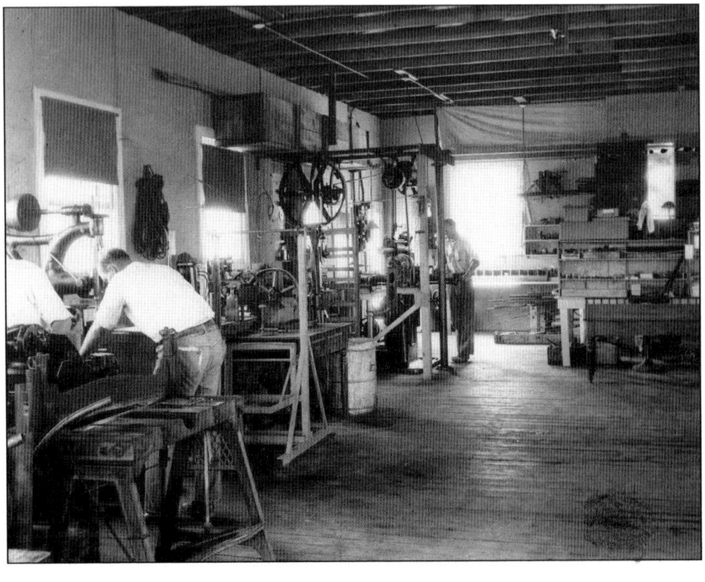

Goddard carefully designed his workshop to house the essential equipment required for constructing and assembling rocket components. Inside, he installed saws, lathes, drills, and a variety of specialized tools, creating a space tailored to the precise demands of his experimental work. The close proximity to his home meant he was free from the delays of commuting, allowing him to dedicate long hours to refining designs and fabricating and testing parts.

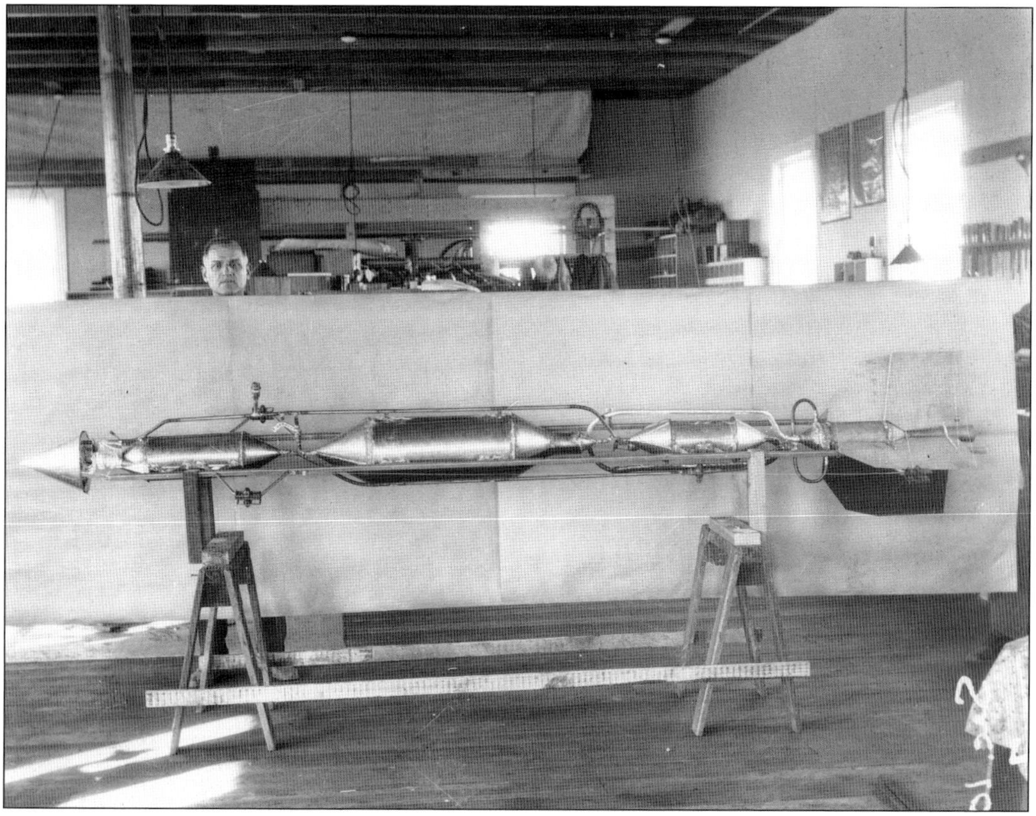

Inside the workshop, assistant Henry Sachs stands behind the first rocket that Goddard would successfully launch from Roswell (and fifth overall). The rocket assemblage stretched 11 feet in length and weighed nearly 34 pounds empty. It included a gas pressure tank that forced the gasoline and oxygen propellants into the combustion chamber, enabling controlled ignition.

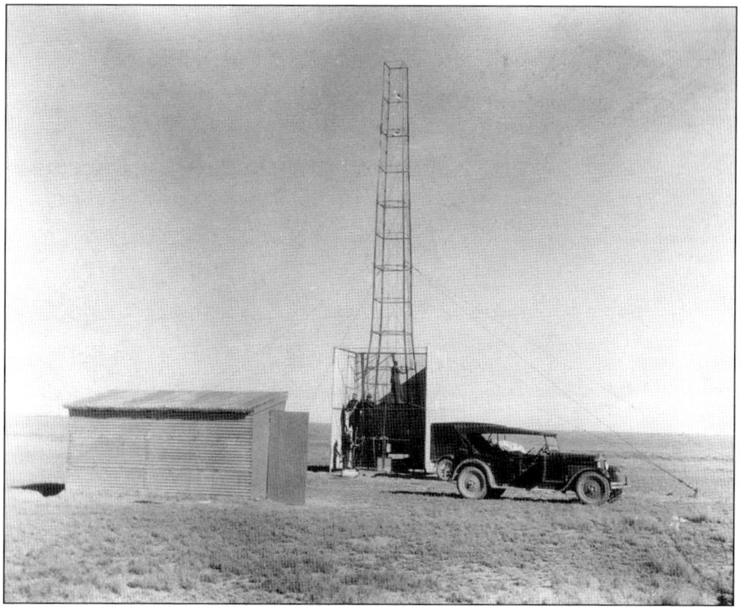

For rocket launches, Goddard found an ideal spot about a dozen miles from Mescalero Ranch in an area called Eden Valley, which was located on some property owned by rancher Oscar White. Goddard and his team reassembled and modified the well-traveled 60-foot tower here and installed an observing shelter nearby. Between 1930 and 1941, Goddard successfully launched 31 rockets from here.

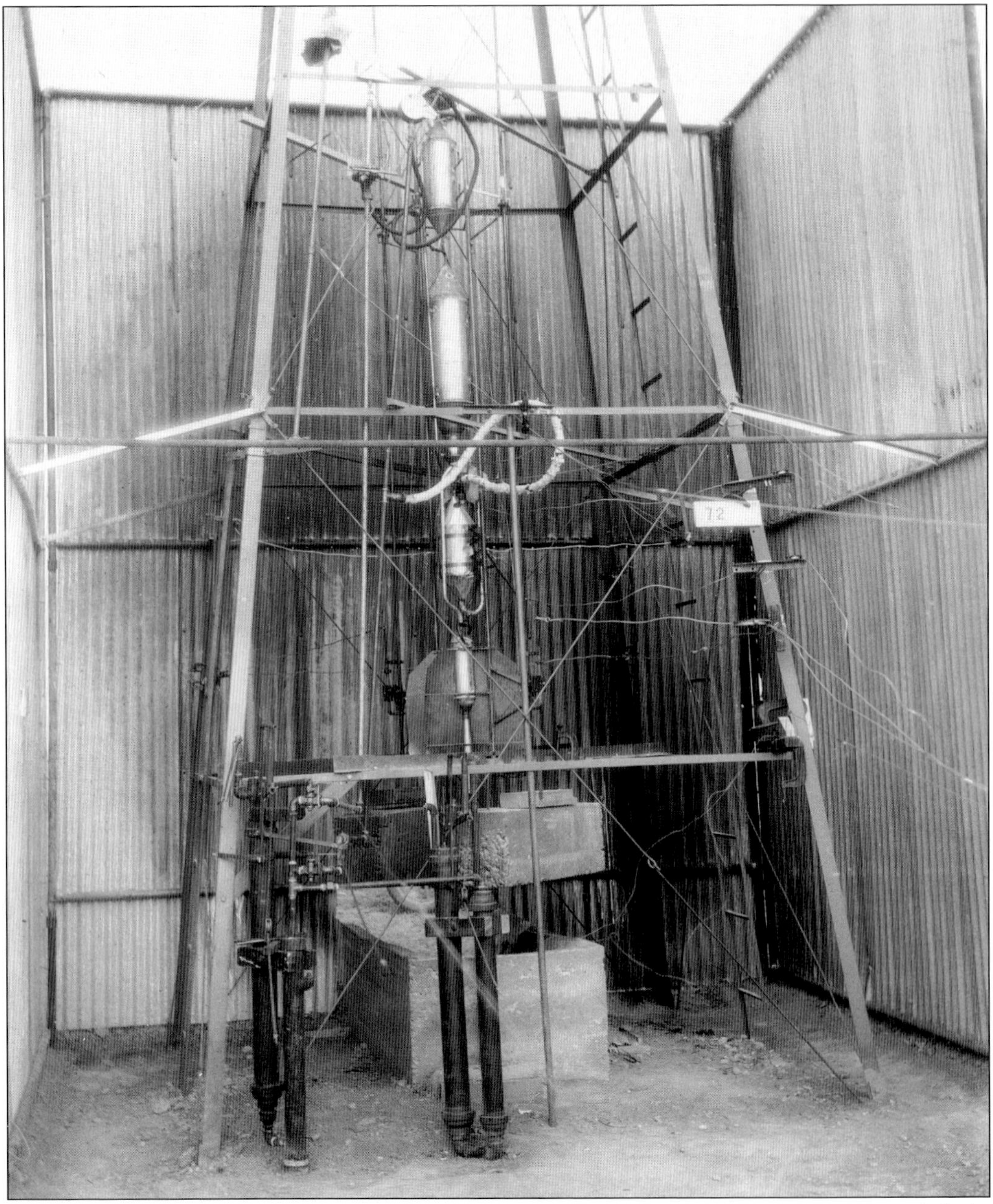

Robert Goddard used this rocket, launched on December 30, 1930, for his first successful flight from Roswell, New Mexico. Drawing on lessons from earlier launches, static tests, and other experiments in Massachusetts, he incorporated countless refinements. The result was a dramatic leap in performance: the rocket reached a speed of 500 miles per hour, soared to an altitude of 2,000 feet, and landed approximately 1,000 feet from the launch site. Although its parachute did not fully deploy, it slowed the descent enough to prevent damage upon impact—a testament to the improved design and engineering. True to his methodical nature, Goddard recorded the flight's details in his diary, noting the technical outcomes and observations. He ended the entry with a personal touch, describing how he and his wife, Esther, celebrated the achievement: "Went to Princess Theater with E., in evening, and saw *Billy the Kid*."

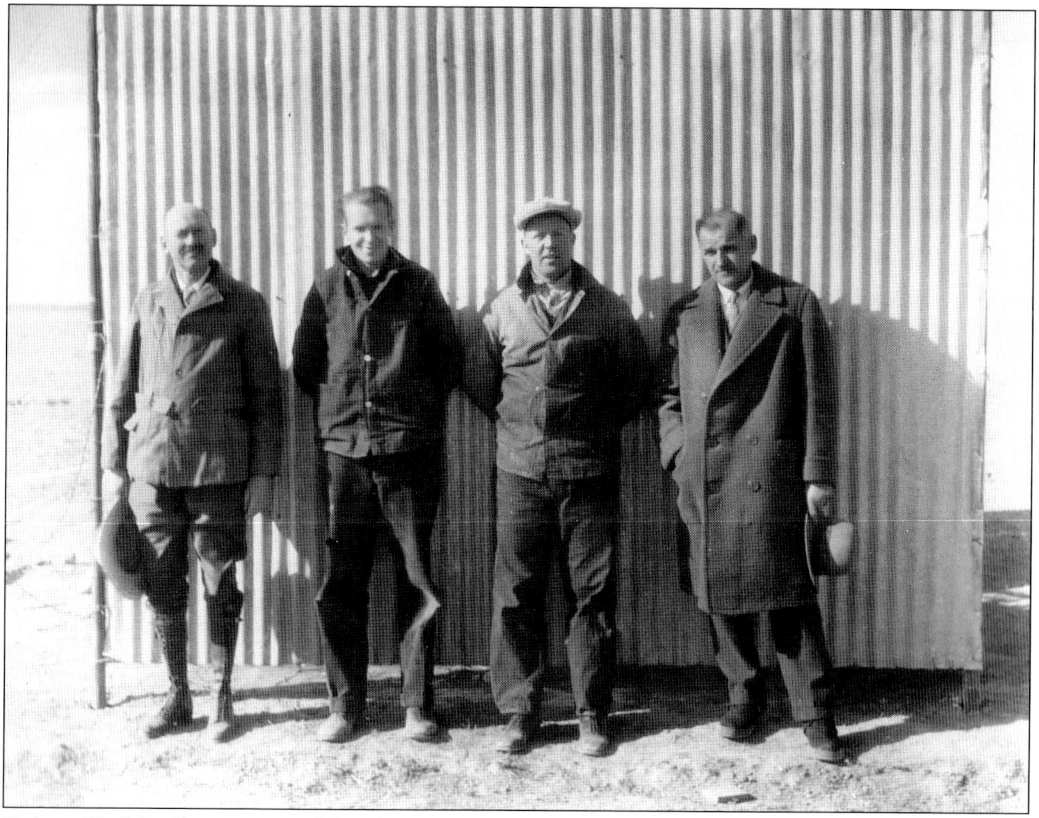

Robert Goddard's team posed for this photograph during their first year in Roswell. At far left stands Goddard, affectionately called "Boss" by his crew. Next is Charles Mansur, a welder and brother of former assistant Larry Mansur. Beside him is Albert Kisk, Goddard's brother-in-law and also a welder. At the end is Henry Sachs, the team's instrument maker, whose precision work was essential to the success of Goddard's pioneering rocket experiments.

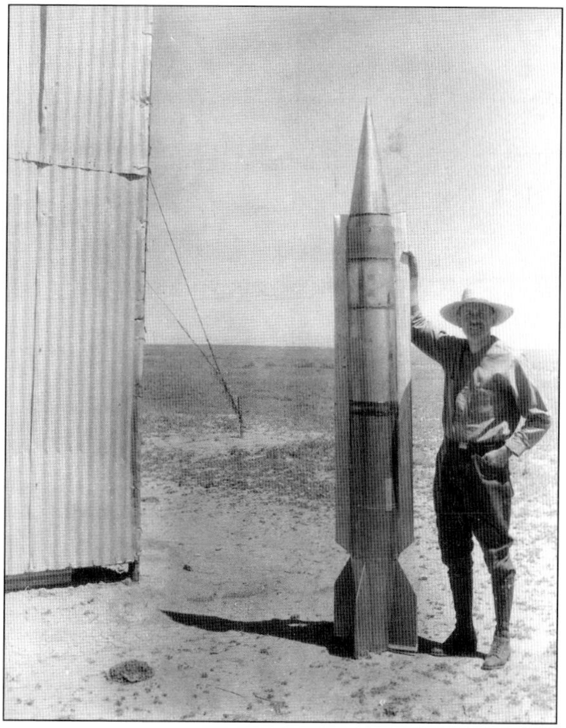

Goddard stands beside the rocket he launched on September 29, 1931, a nine-foot-long design that marked a significant advancement in his work. For the first time, he enclosed the rocket's components in a protective jacket and added a streamlined casing, also known as a fairing or nose cone. These features improved aerodynamic flow and shielded internal parts during flight. Goddard noted in his diary that the rocket's ascent was "like a fish swimming upward."

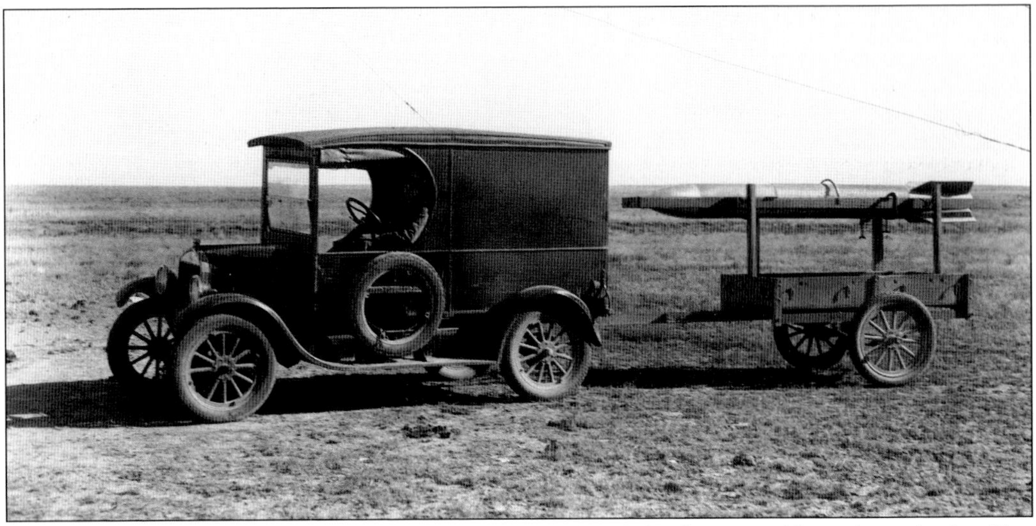

Although Robert Goddard frequently refined and updated the design and launch methods of his rockets, some aspects of his operations remained remarkably consistent over the years. One enduring routine was the way he transported rockets from the workshop at Mescalero Ranch to the launch site in Eden Valley. Despite technological advancements, the process still involved loading each rocket onto a simple trailer and towing it behind a vehicle.

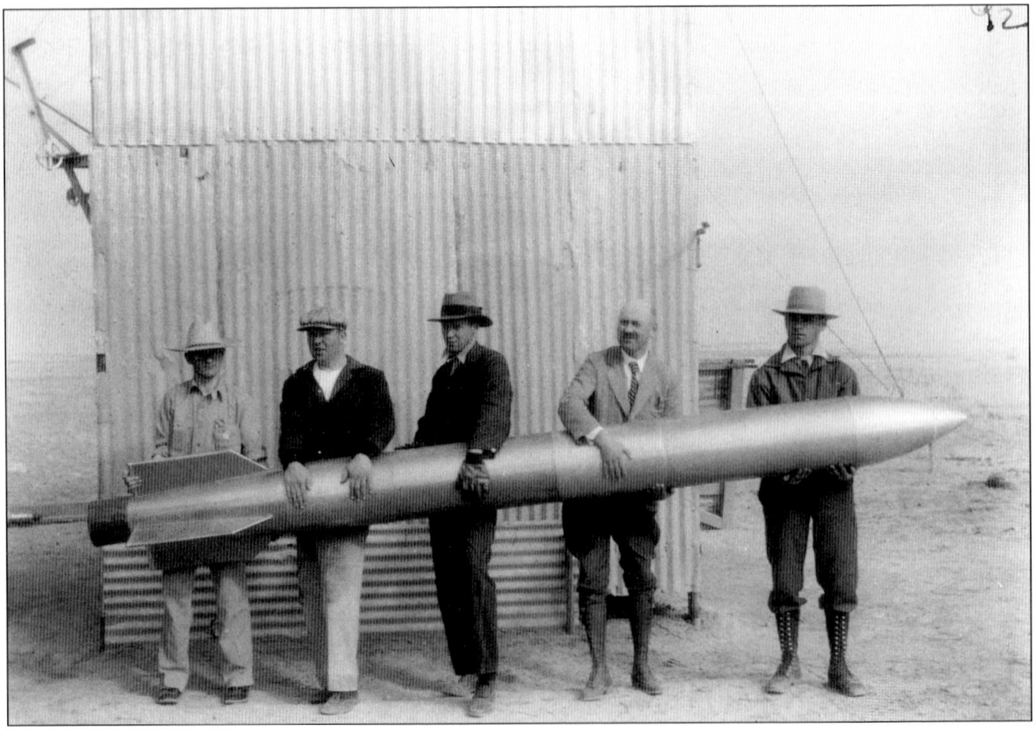

Goddard and his team hold a rocket that they launched on April 13, 1932. It featured gyroscopic-controlled vanes—his first attempt at using this technology to guide flight. It marked a major innovation but also the end of his initial research at Roswell. With the Guggenheim Foundation funding cut due to the Great Depression, this was the last Roswell launch for two years. In June, Goddard, his crew, and Esther packed up their equipment and returned to Worcester.

Robert Goddard and his team returned to Rowell in 1934 as Guggenheim funding was reinstated. By the following year, his work in Roswell had reached full momentum. That year alone, he successfully launched seven rockets, each representing a leap forward in design and performance. In addition to flight tests, Goddard conducted numerous static trials and experimental procedures. In this photograph, Goddard uses an electronic control panel to operate a rocket.

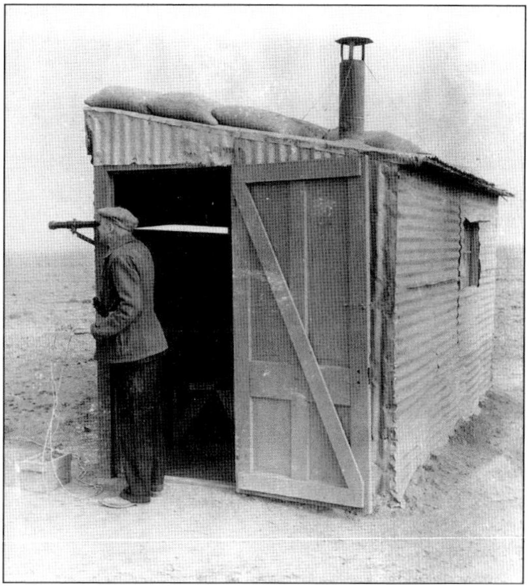

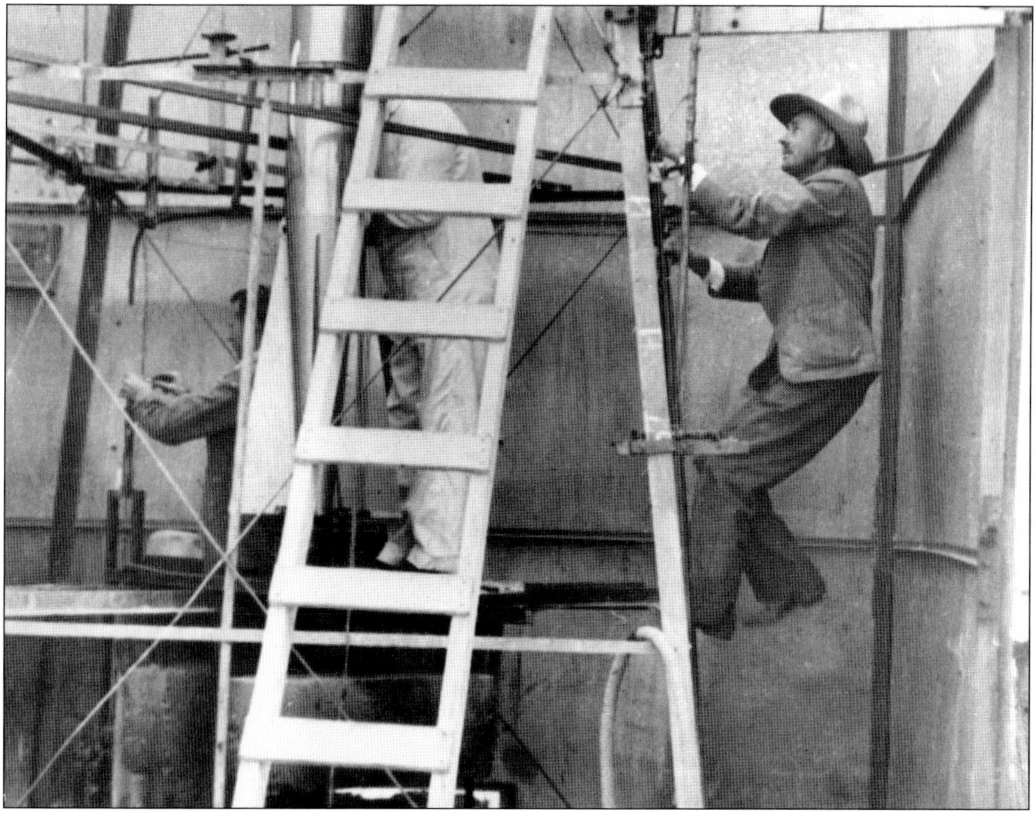

Goddard here joins his crew in the launch tower on September 23, 1935, preparing for a rocket test that ultimately failed due to an igniter malfunction. Despite the setback, Goddard remained optimistic, viewing such failures as valuable learning opportunities. Each misstep provided him with the insight to modify and improve future designs. His patience and commitment to rigid experimentation were key to his pioneering rocketry research.

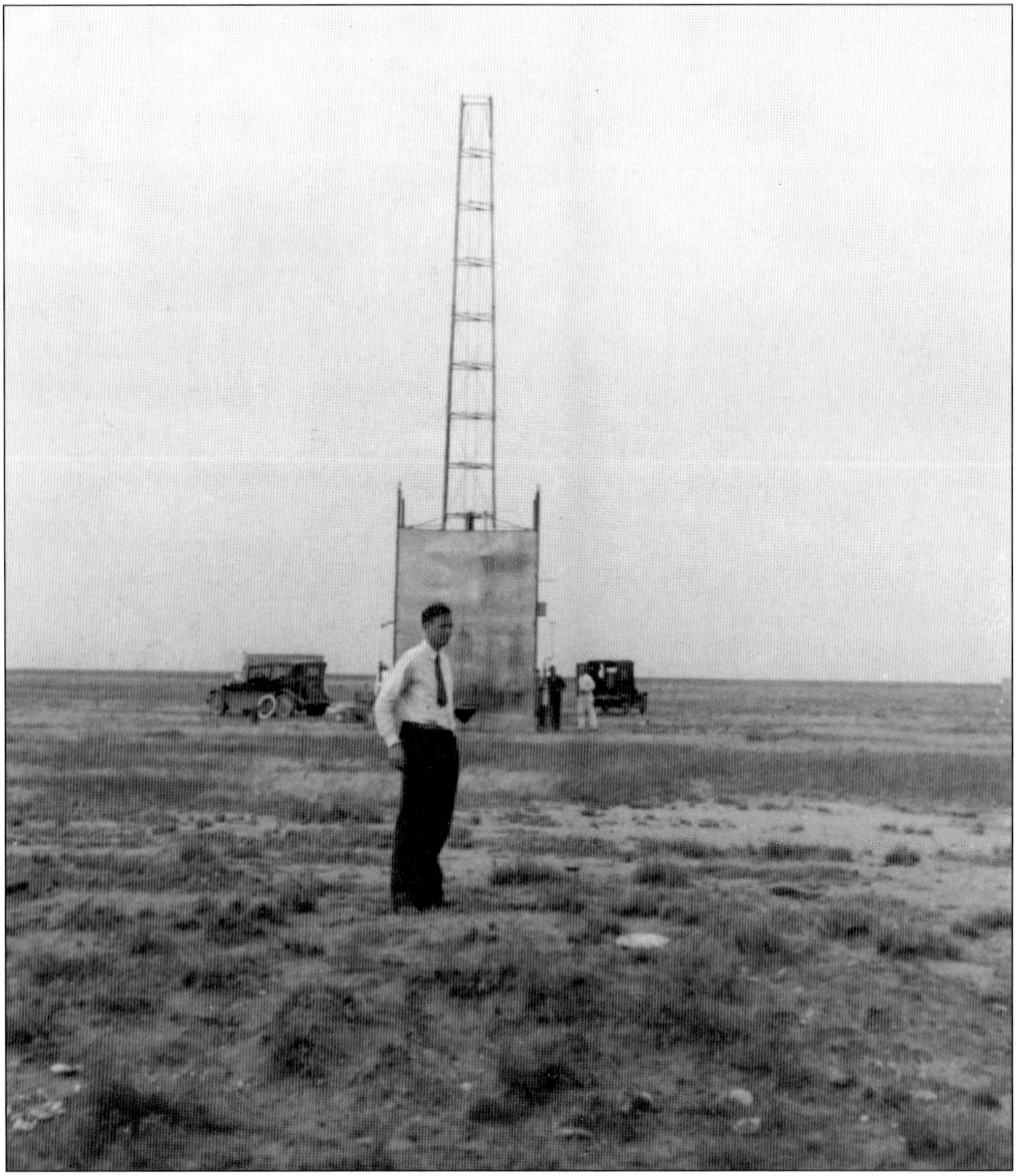

Robert Goddard welcomed some special guests to Roswell on September 22, 1935—Charles Lindbergh (shown here in front of the launch tower), Harry Guggenheim, and Guggenheim's wife, Carol. Eight years after his epic solo flight across the Atlantic Ocean that inspired people around the globe and sparked widespread interest in commercial air travel, Lindbergh was still one of the most famous people in the world. This fame allowed him to wield a lot of influence, and this was instrumental in Goddard receiving funding from the Guggenheim Foundation. Harry Guggenheim had assumed leadership of the Guggenheim Foundation following the 1930 death of his father, Daniel. The Guggenheims and Lindbergh stayed until September 25, eager to witness a launch and assess the progress of Goddard's research firsthand. Unfortunately, due to poor weather and technical issues, they never got to see a launch. Nevertheless, both Lindbergh and Guggenheim were impressed by the breadth and promise of Goddard's experiments. Shortly afterward, Guggenheim announced that the foundation would continue funding Goddard's research.

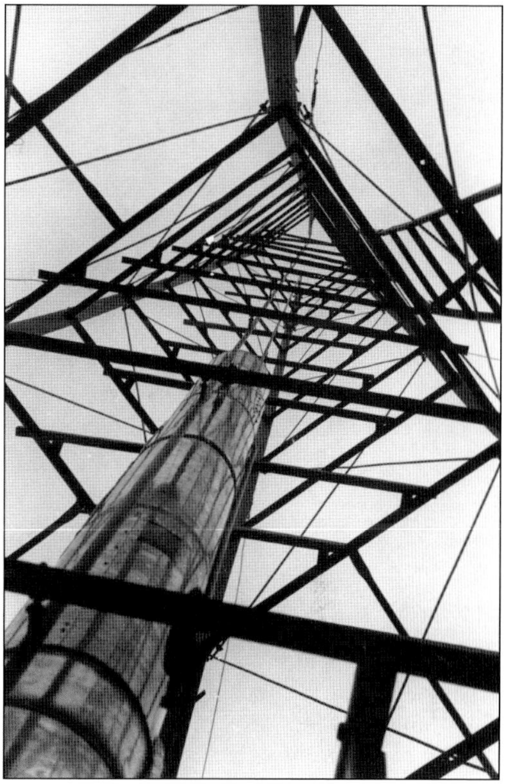

Although Lindbergh and the Guggenheims did not see a rocket launch during their September 1935 visit to Roswell, Lindbergh made the most of the experience by capturing two striking photographs of the launch tower. The image at right presents a dramatic upward view from the base of the tower, with the rocket extending toward the sky. The photograph below, taken from the upper portion of the tower, looks down toward the ground, providing a unique perspective of the launch site's structure and surroundings. These images document the physical structure of Goddard's testing site while offering a glimpse of Lindbergh's eye for detail.

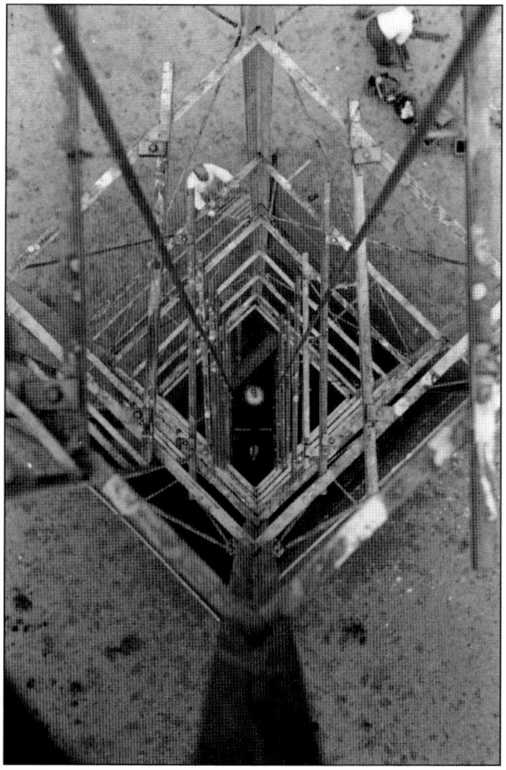

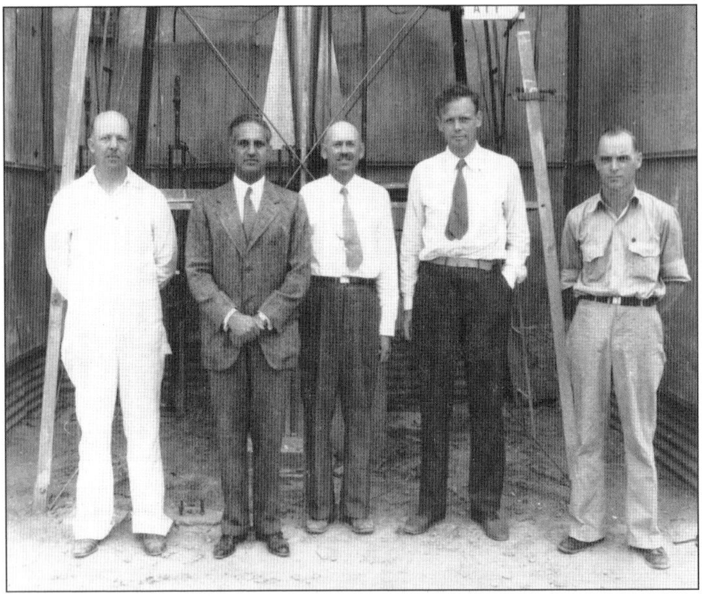

Captured in a group photograph during the September 1935 visit by Lindbergh and the Guggenheims are, from left to right, Albert Kisk, Harry Guggenheim, Robert Goddard, Lindbergh, and Ole Ljungquist. The image reflects a compelling blend of personalities: a visionary scientist, a legendary aviator, a powerful patron, and the dedicated crew whose names may be lesser known but whose behind-the-scenes contributions were vital to the progress of science.

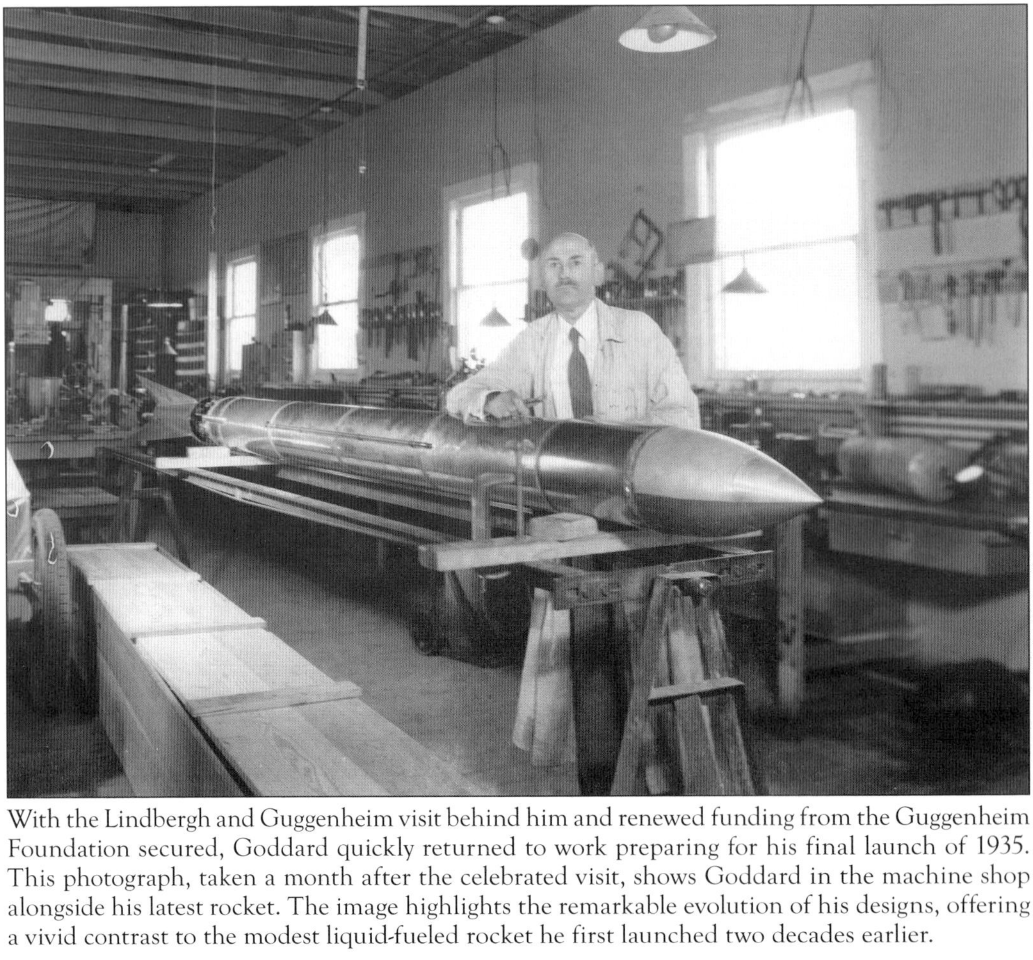

With the Lindbergh and Guggenheim visit behind him and renewed funding from the Guggenheim Foundation secured, Goddard quickly returned to work preparing for his final launch of 1935. This photograph, taken a month after the celebrated visit, shows Goddard in the machine shop alongside his latest rocket. The image highlights the remarkable evolution of his designs, offering a vivid contrast to the modest liquid-fueled rocket he first launched two decades earlier.

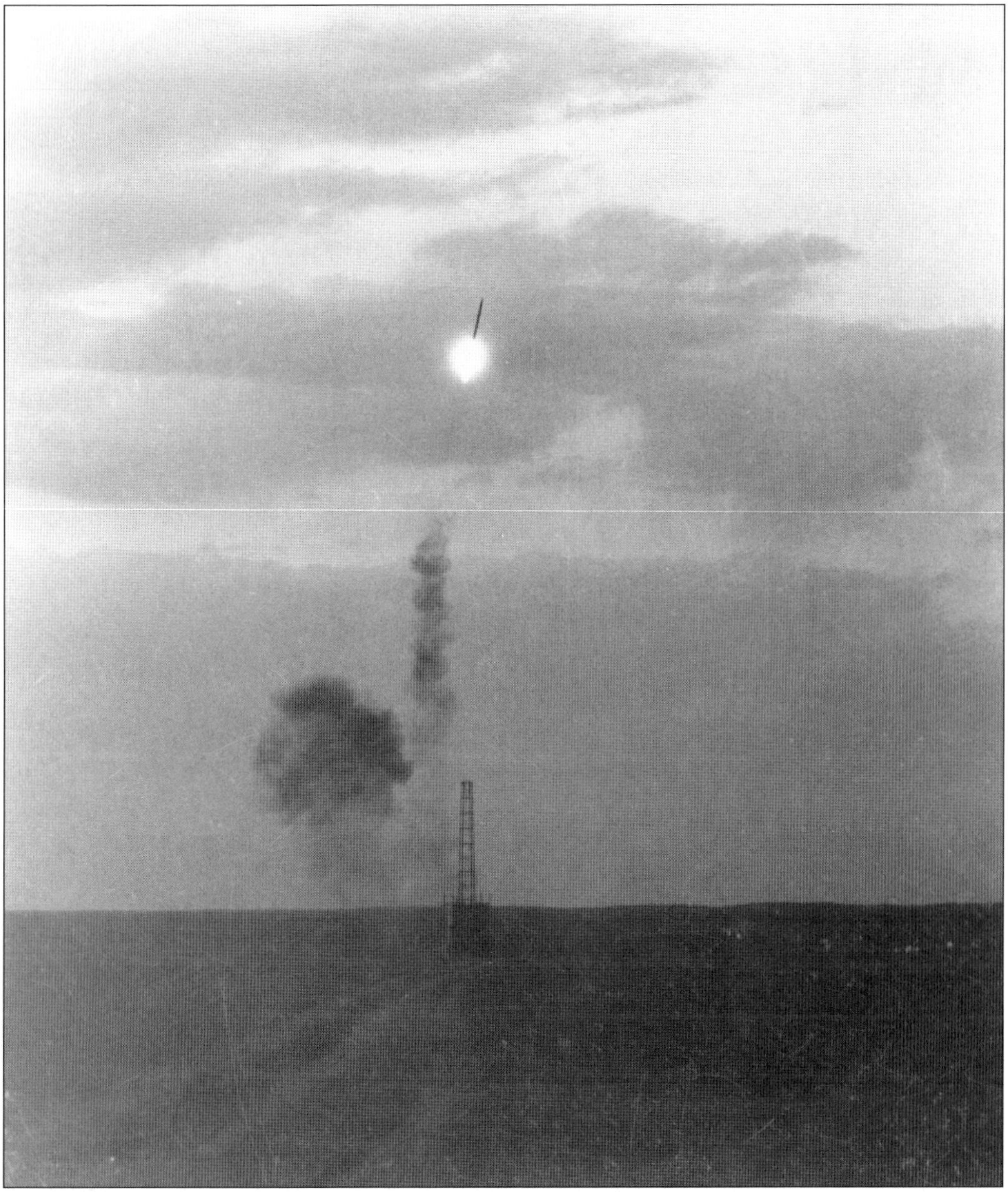

In the early years of his liquid-fueled rocket experiments, Robert Goddard concentrated on one primary goal: getting the rockets off the ground. These initial launches focused on achieving liftoff and evaluating potential propulsion systems. Once Goddard solved these issues, he began exploring more complex challenges, such as flight stability. Maintaining a consistent, vertical course proved difficult, and he spent years refining his designs and characteristically testing, testing, and testing to address the issue. By the mid-1930s, his persistence began to pay off. On May 19, 1937, just after sunset, he achieved his most successful and stable flight to date. The rocket climbed to an altitude of 3,250 feet and remained nearly vertical throughout its ascent, marking a major breakthrough in his quest for controlled flight. This image captures the rocket in mid-flight and is a testament to how far Goddard's work had progressed since his earliest launches on Aunt Effie's farm two decades earlier, when his rockets stayed aloft for mere seconds and swerved uncontrollably in all directions.

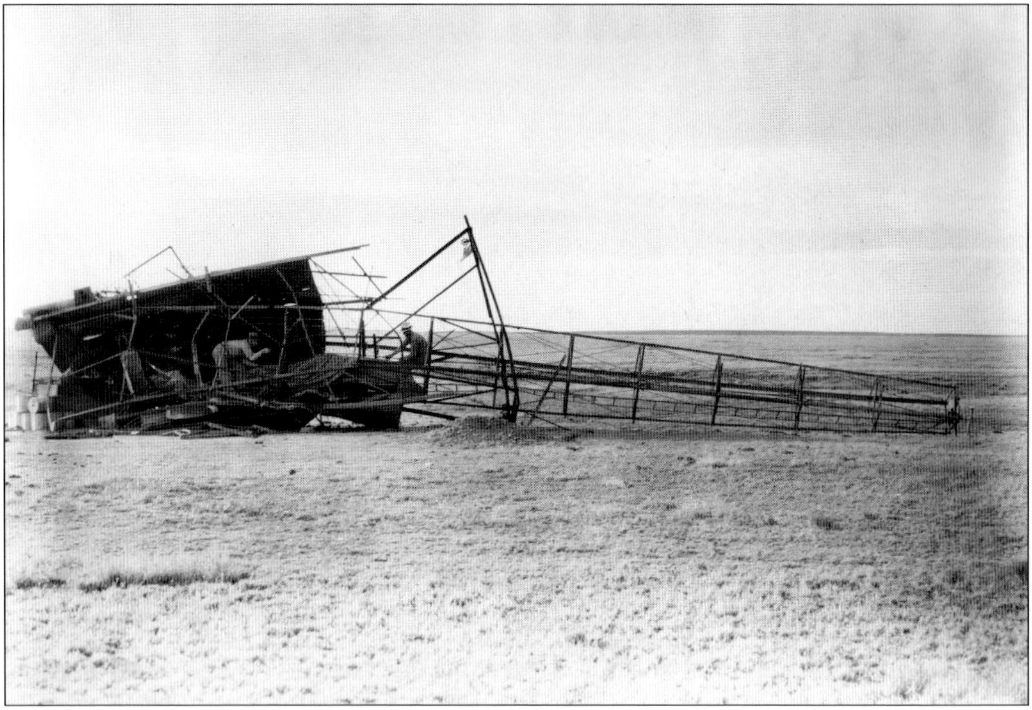

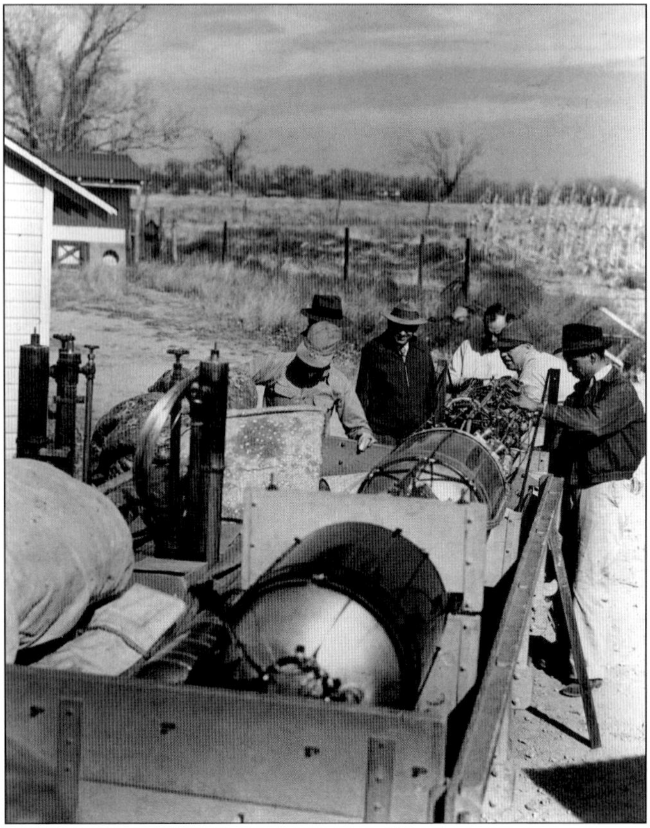

In his decades of rocket research, Goddard had overcome a wide variety of challenges. Just when he might have thought he'd seen it all, he experienced a new one on June 15, 1938. That afternoon, a tornado touched down at the launch site. The launch tower took a direct hit, turning it into, as Goddard described in his diary, "a pile of rods and bent angles."

In this 1940 photograph, Robert Goddard (third from left) observes his crew as they load a rocket, stripped of its outer shell, onto a trailer for transport from the machine shop to the Eden Valley launch site. This model was the first in which Goddard employed turbopumps to feed propellants into the combustion chamber, replacing his earlier pressure-based gas system.

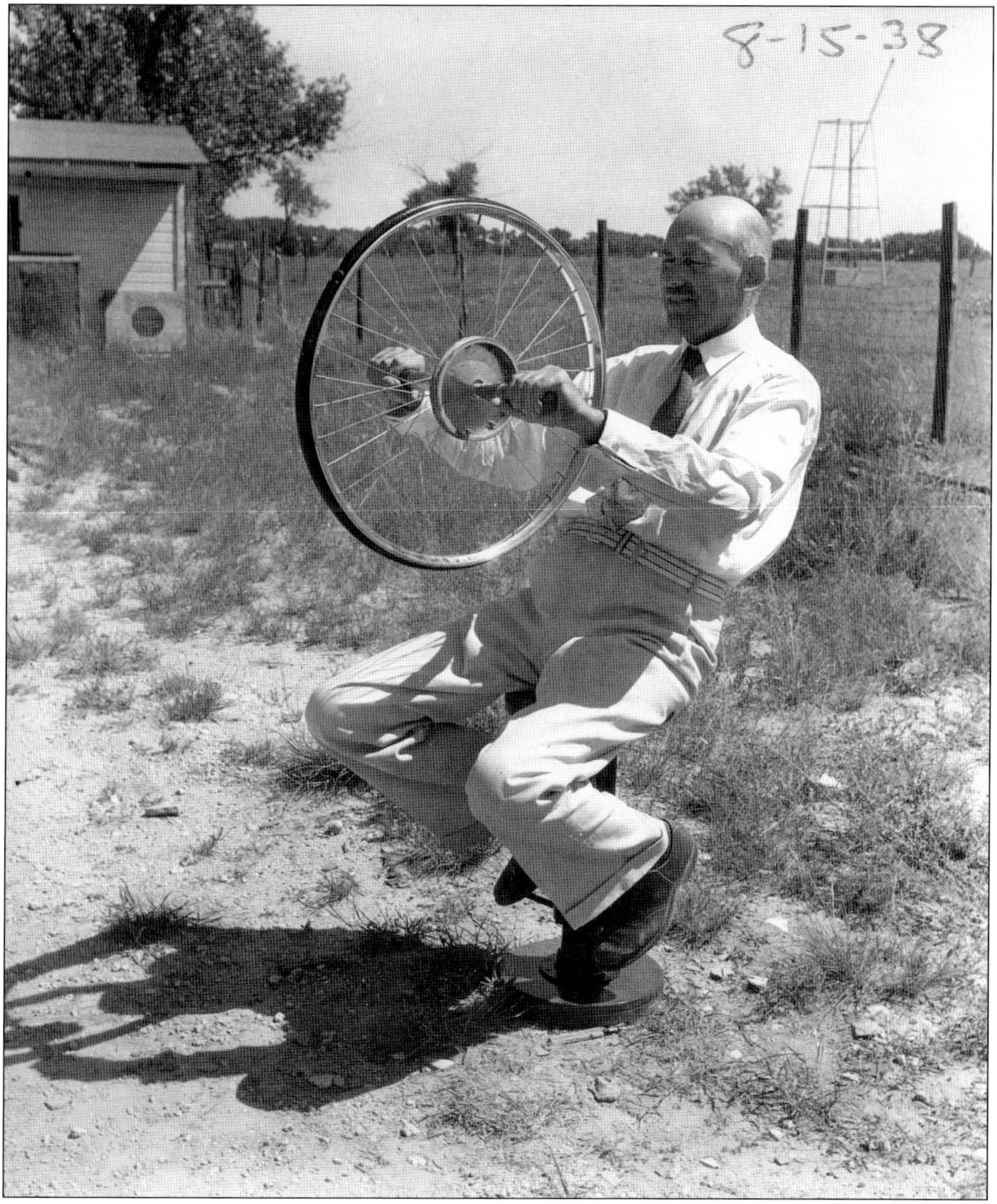

Goddard toiled for years to find ways to control his rockets in flight. When he applied for a patent involving gyroscopic rocket guidance, he built a bicycle gyrostat, a device consisting of a casing that surrounds a spinning rotor. He used this to demonstrate principles of angular momentum and how gyroscopes could stabilize and steer a rocket. This picture shows Goddard holding an actual bicycle wheel, but the device he built consisted of a wooden bicycle wheel that he modified by cutting notches into it. He coupled it with metal piping modified to simulate the gyroscopic activity. The gyrostat epitomized Goddard's approach of not just theorizing but doing the hands-on work of building and testing. He used this approach throughout his entire life, from his childhood days of building a frog hatchery prototype to his revolutionary liquid-fueled rockets. Though crude, the gyrostat device helped lay the foundation for navigation systems used in modern aerospace.

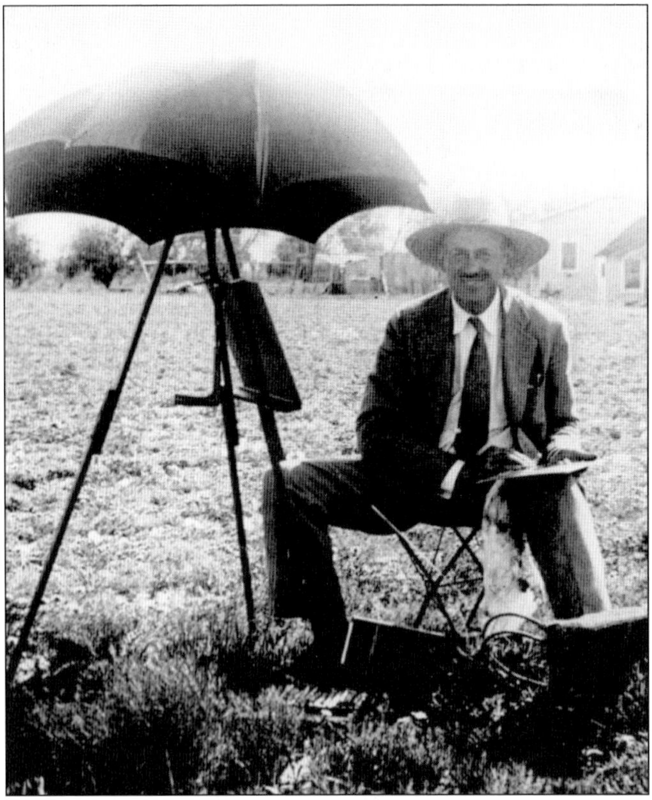

Though remembered primarily as a pioneering rocket scientist, Robert Goddard cultivated a range of personal interests that brought him relaxation and balance. He loved playing the piano so much that he transported his own piano from Worcester to Roswell so he could play whenever inspiration struck. In addition to music, he also had a lifelong fascination with drawing and took up painting.

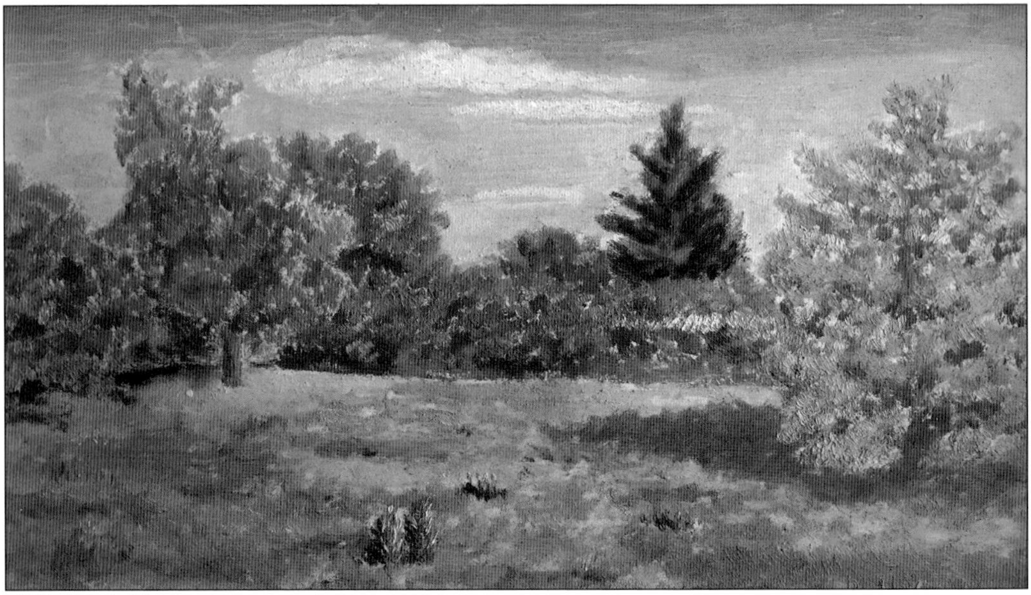

Goddard especially enjoyed painting landscapes, from the sweeping vistas of New Mexico to the lush countryside of his Worcester hometown, as shown in this piece titled *Near Hope Cemetery*. This painting is one of many Goddard historical pieces housed at the Clark University library named in his honor. Other significant repositories of Goddard historical material include Worcester Polytechnic Institute, the Smithsonian Air & Space Museum, and the Roswell Museum.

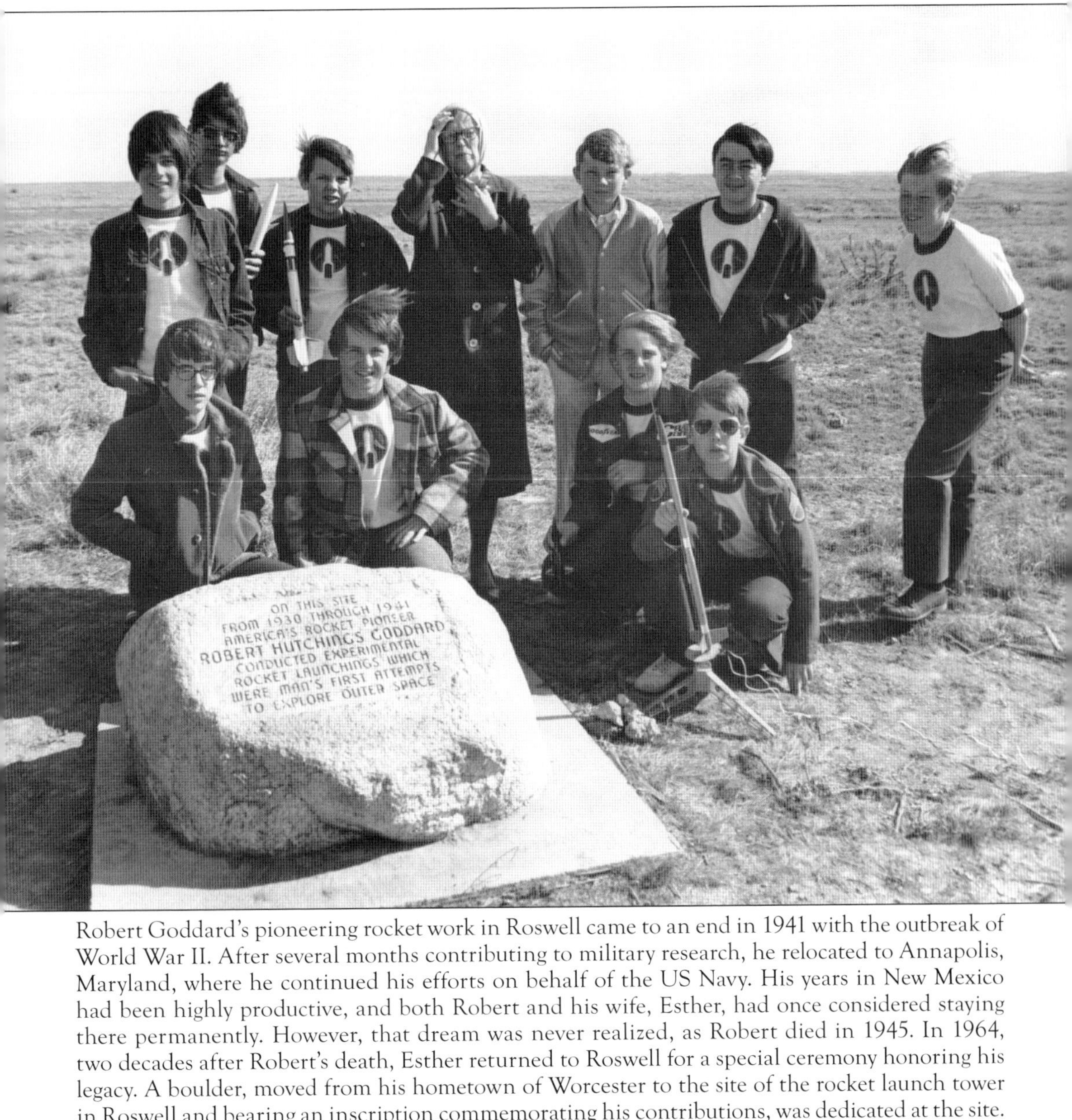

Robert Goddard's pioneering rocket work in Roswell came to an end in 1941 with the outbreak of World War II. After several months contributing to military research, he relocated to Annapolis, Maryland, where he continued his efforts on behalf of the US Navy. His years in New Mexico had been highly productive, and both Robert and his wife, Esther, had once considered staying there permanently. However, that dream was never realized, as Robert died in 1945. In 1964, two decades after Robert's death, Esther returned to Roswell for a special ceremony honoring his legacy. A boulder, moved from his hometown of Worcester to the site of the rocket launch tower in Roswell and bearing an inscription commemorating his contributions, was dedicated at the site. In this photograph from that day, Esther is surrounded by a group of young rocketry enthusiasts, a fitting tribute to the enduring impact of her husband's work. A statue of Robert Goddard, one of his rocket frames, and other displays and artifacts in Roswell celebrate the rocket scientist's critical work there.

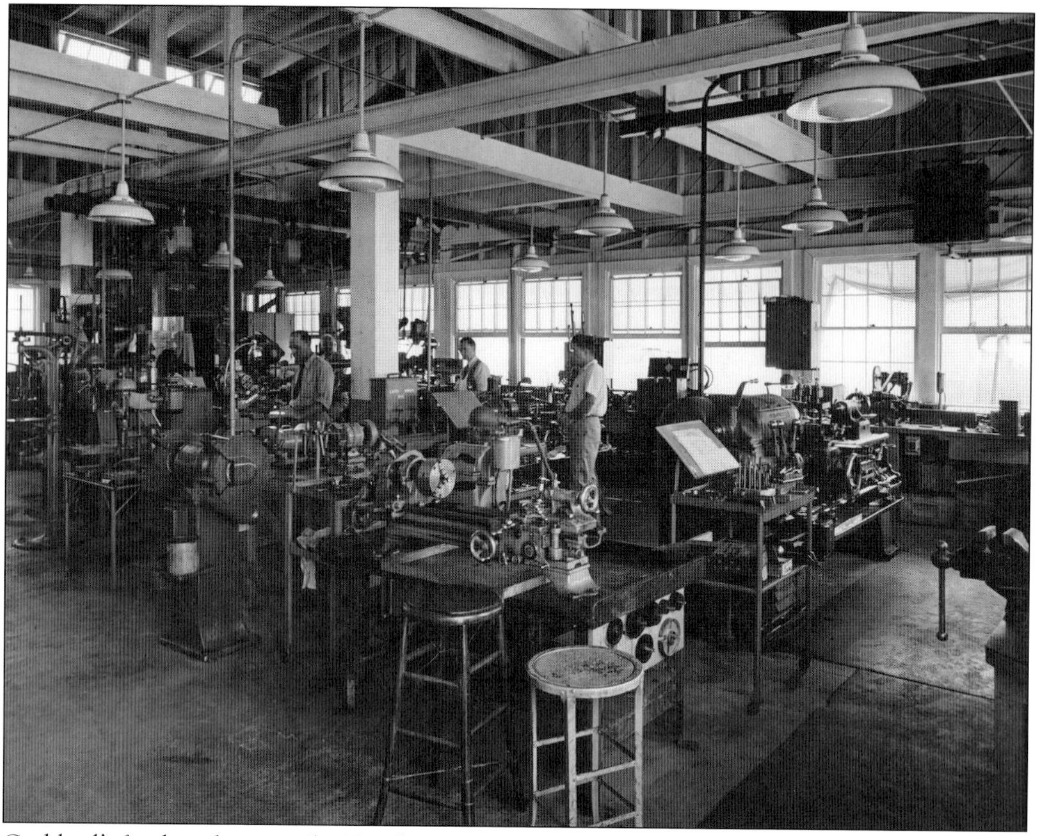

Goddard's final work was at the Naval Engineering Experiment Station in Annapolis, Maryland, where he developed JATO rockets. These were designed to help aircraft launch more effectively, especially when carrying heavy payloads or operating from short runways. Goddard's work helped advance military aviation during World War II.

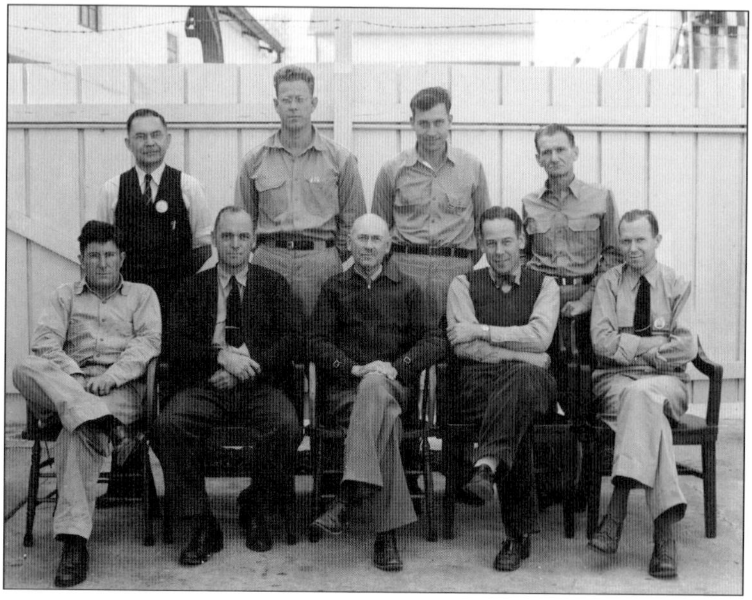

Robert Goddard is shown here toward the end of his life with his staff at the Naval Engineering Experiment Station in Annapolis, Maryland. His early-life bout with tuberculosis, compounded by his lifelong habit of smoking cigars, caught up with him in 1945 while working here. He was diagnosed with throat cancer and died on August 10 at the age of 62. Esther would survive him by nearly four decades.

Six

Worcester

Legacy

During his life, he was little known and less appreciated,
yet his efforts had incalculable influence on the course of history.

—Wernher von Braun and Frederick I. Ordway,
History of Rocketry and Space Travel, 1966, reprinted 1969

Few individuals have shaped the future of humankind as profoundly as Robert Goddard, yet his legacy remains most deeply rooted in Worcester, Massachusetts. This connection is not only because Worcester was his hometown but also because it was the place where he made many of his most groundbreaking contributions to science and technology. Equally important, however, is the role played by his wife, Esther, who devoted the decades following his death to ensuring his work was remembered and celebrated.

In the later years of their life together, Robert and Esther considered settling permanently in Roswell. That plan might have materialized if not for the outbreak of World War II, which disrupted the lives of millions. Though nearly 60 years old when the war began, Robert felt compelled to contribute to the war effort and accepted a position with the US Navy in Maryland, where he worked on propulsion systems.

When Robert died of cancer in 1945, Esther chose not to remain in Maryland. Instead, she returned to Worcester, where she was surrounded by family, friends, and familiar places. Back at the Goddard home, she found not only comfort but also purpose. She took on the role of archivist, advocate, and steward of her husband's legacy. She meticulously transcribed his diaries and technical writings, worked to have his patents formally recognized, and campaigned for his contributions to be acknowledged by the broader scientific community.

Her efforts led to significant honors, including the naming of the NASA Goddard Space Flight Center in his memory. Esther also participated in dozens of ceremonies, lectures, and commemorative events, often serving as the public face of Robert's legacy. Through her tireless dedication, she ensured that his pioneering work in rocketry would not be forgotten and that Worcester would remain central to the story of one of the most visionary minds of the 20th century.

Esther's commitment transformed personal grief into a lifelong mission, one that preserved Robert Goddard's place in history and inspired generations to follow their dreams.

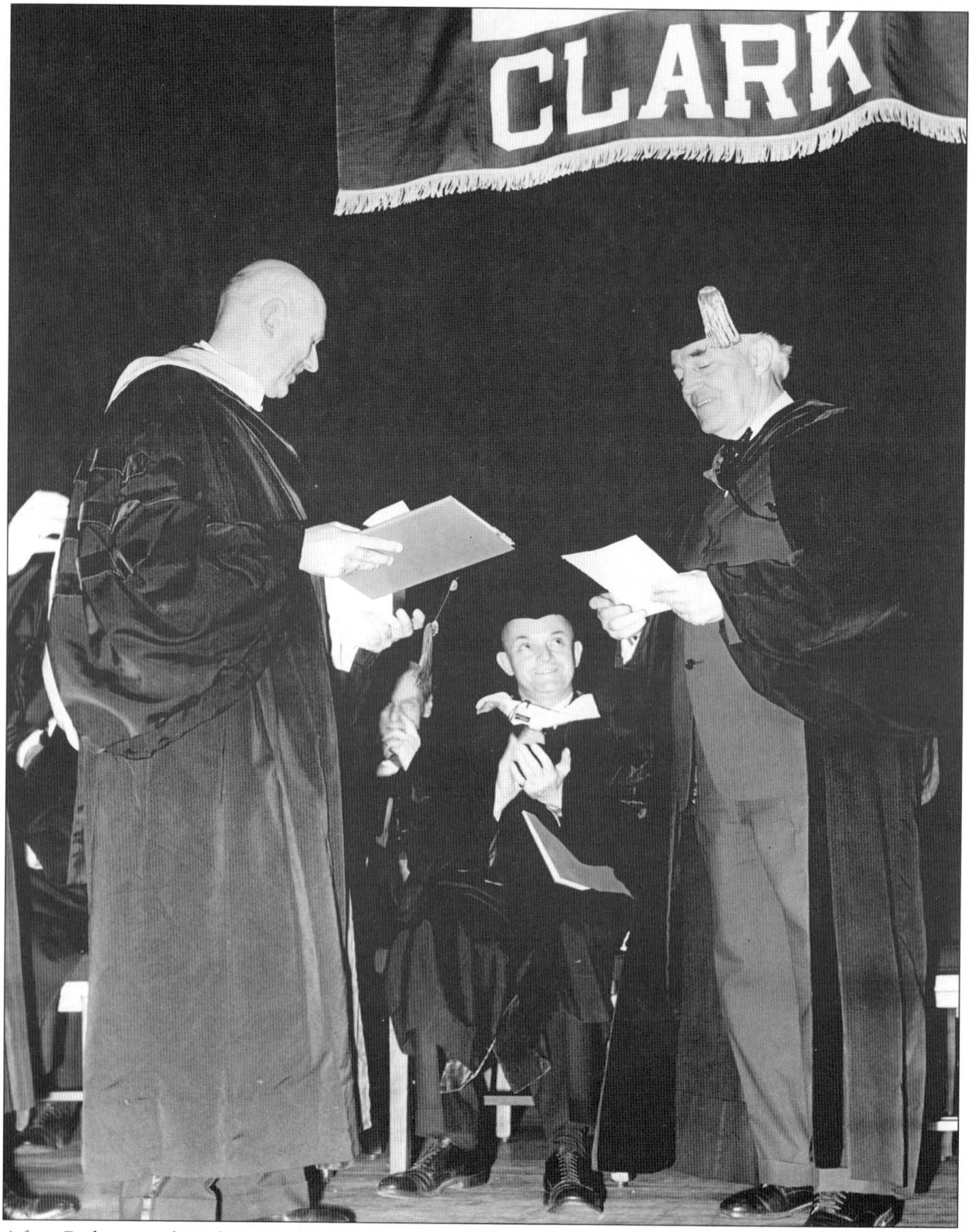

After Robert and Esther Goddard moved away from Worcester—first to New Mexico, later to Maryland—they regularly returned to Worcester to visit friends and family and to attend special events. One particularly meaningful visit occurred in 1945, when Clark University, where Robert had earned his doctorate and served as a professor, awarded him an honorary doctor of science degree in recognition of his pioneering work in rocketry. The ceremony was held on June 2, just weeks before his death on August 10, 1945.

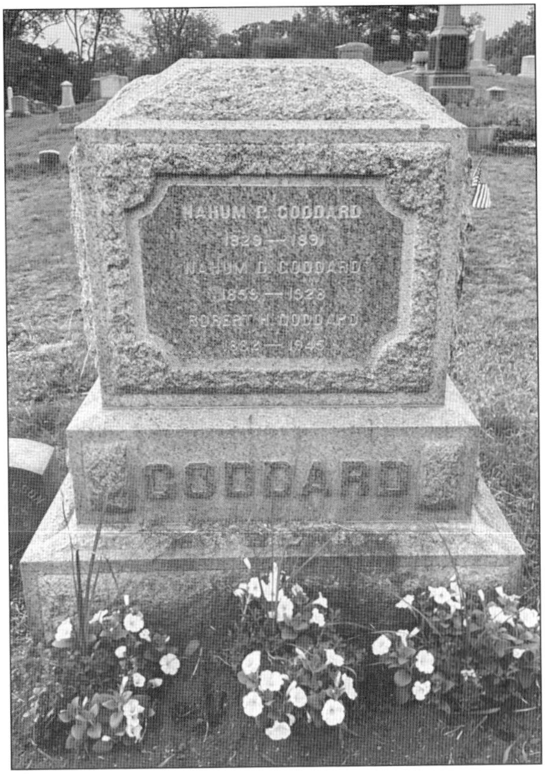

Despite his early battle with tuberculosis and other health challenges that left his body fragile, Robert Goddard lived with remarkable intensity. He worked tirelessly on rocket development, and when he needed to unwind, he enjoyed smoking cigars and sipping martinis—small indulgences that punctuated long days of work. By 1945, while working for the US Navy in Maryland, Goddard's mind remained sharp, but his body faltered. Diagnosed with throat cancer, he died on August 10, 1945, at the age of 62. His body was returned to Worcester and buried in the Goddard family plot at Hope Cemetery. A large central marker bears on the front the names of three generations of Goddard men—Robert, his father, Nahum Danford, and his grandfather Nahum Parks— with his brother, Richard, memorialized on the back. Smaller stones surround the monument; Robert's marker sits in the front row, bookended by that of his father, Nahum, to the left and Esther to the right. It serves as a quiet tribute to a life that launched humanity toward the stars. (Both, Kevin Schindler personal collection.)

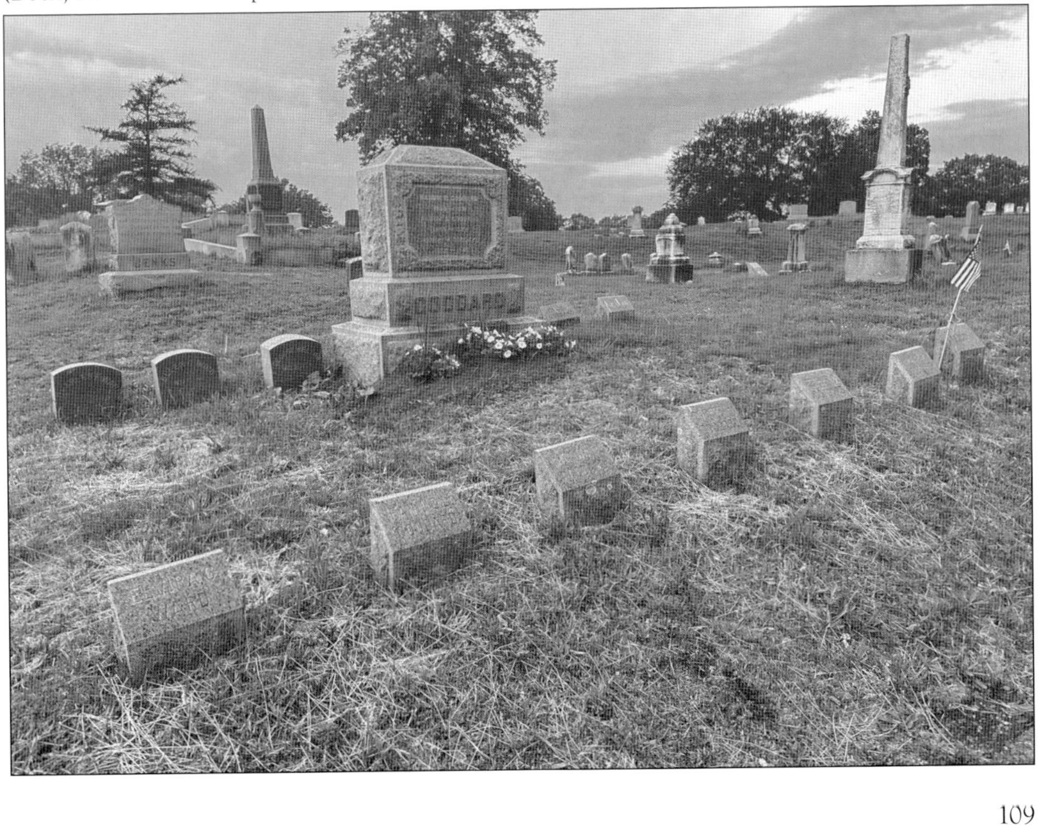

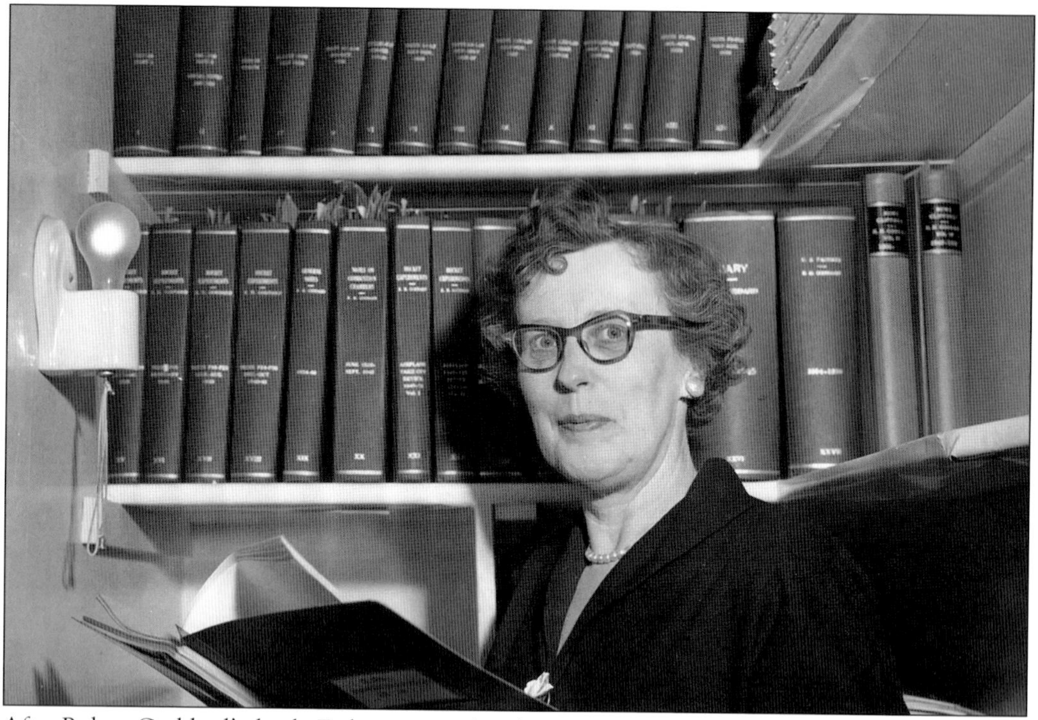

After Robert Goddard's death, Esther returned to their Worcester home and dedicated the remainder of her life to preserving and promoting his legacy. She meticulously organized and published his scientific papers, commissioned a biography to tell his story, and partnered with patent attorney Charles Hawley to secure intellectual property rights for his many innovations. Through her tireless efforts, 131 of Robert Goddard's 214 patents were awarded posthumously.

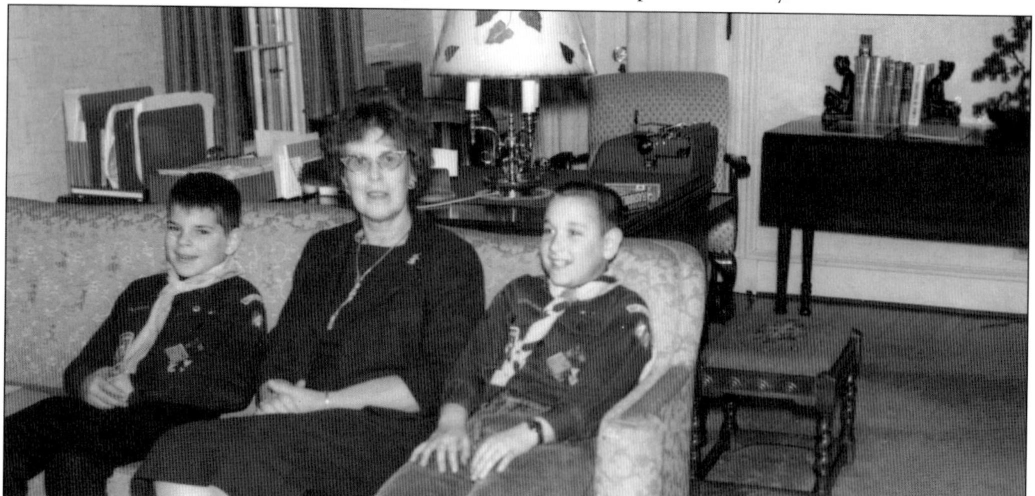

In preserving Robert Goddard's legacy, Esther's efforts involved meeting with community leaders, scientists, and others whose lives had been touched by his pioneering work. But in the spirit of Robert's high school graduation speech, in which he declared, "It is difficult to say what is impossible, for the dream of yesterday is the hope of today and the reality of tomorrow," Esther also made it a point to connect with the dreamers of today and tomorrow: youth. Many young neighborhood children later told stories of being inspired by Esther.

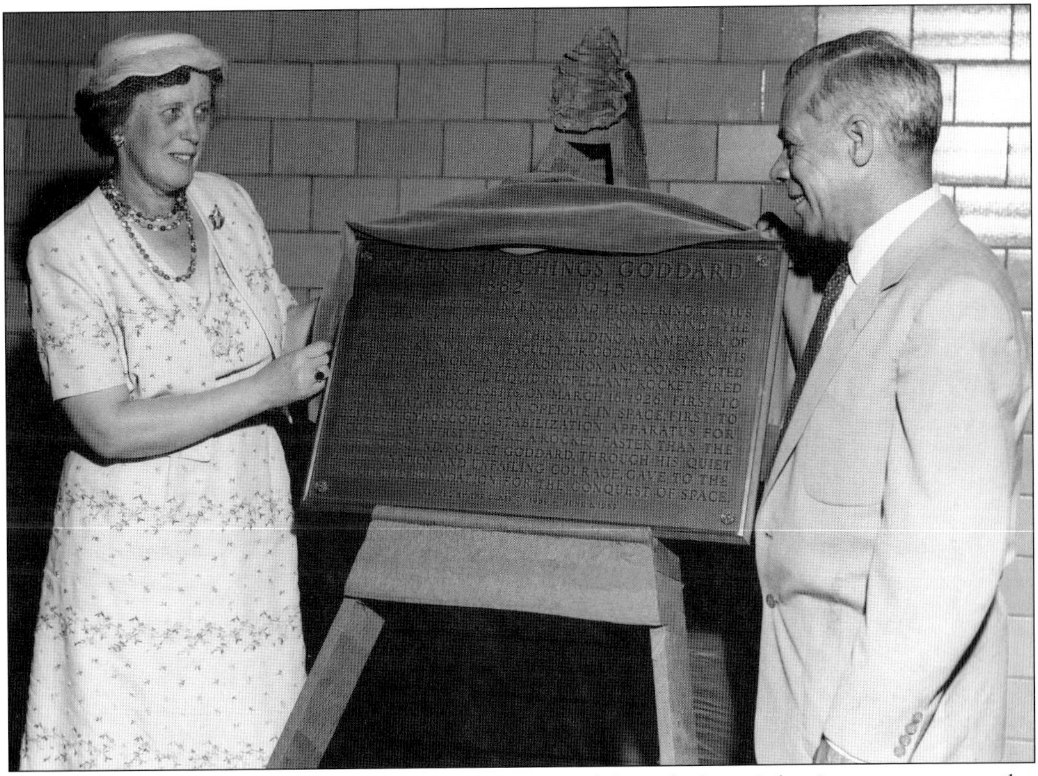

In 1959, Esther joined Clark University trustee and board chair John Jeppson to unveil a commemorative plaque honoring her late husband. Sponsored by the Clark University class of 1959, the plaque was mounted on the outside wall of the mathematics and physics building, where Robert Goddard had conducted many of his most important experiments. The plaque's final line reads, "Robert Goddard, through his quiet determination and unfailing courage, gave to the world the foundation for the conquest of space." It stands as a tribute not only to his scientific achievements but also to his perseverance in the face of skepticism and adversity. The unveiling ceremony was a poignant moment for Esther, who had worked tirelessly to preserve Robert's legacy. Today, the plaque remains a visible reminder of Goddard's enduring impact on science and space exploration. (Above, Robert H. Goddard Library, Clark University; below, Kevin Schindler personal collection.)

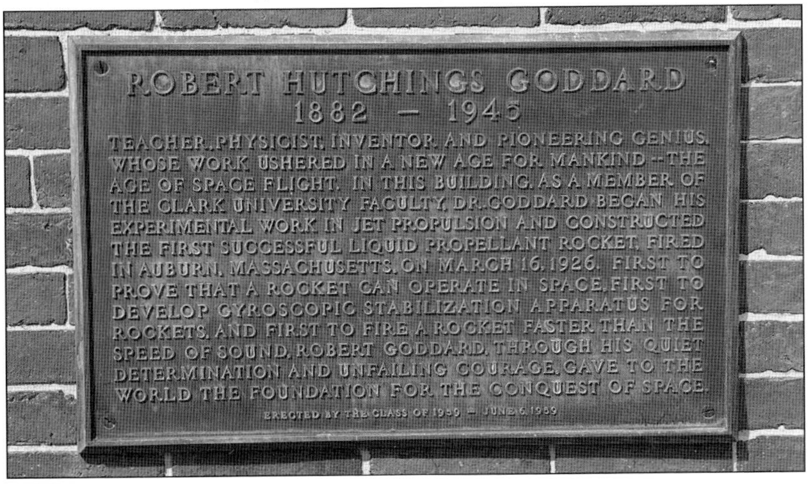

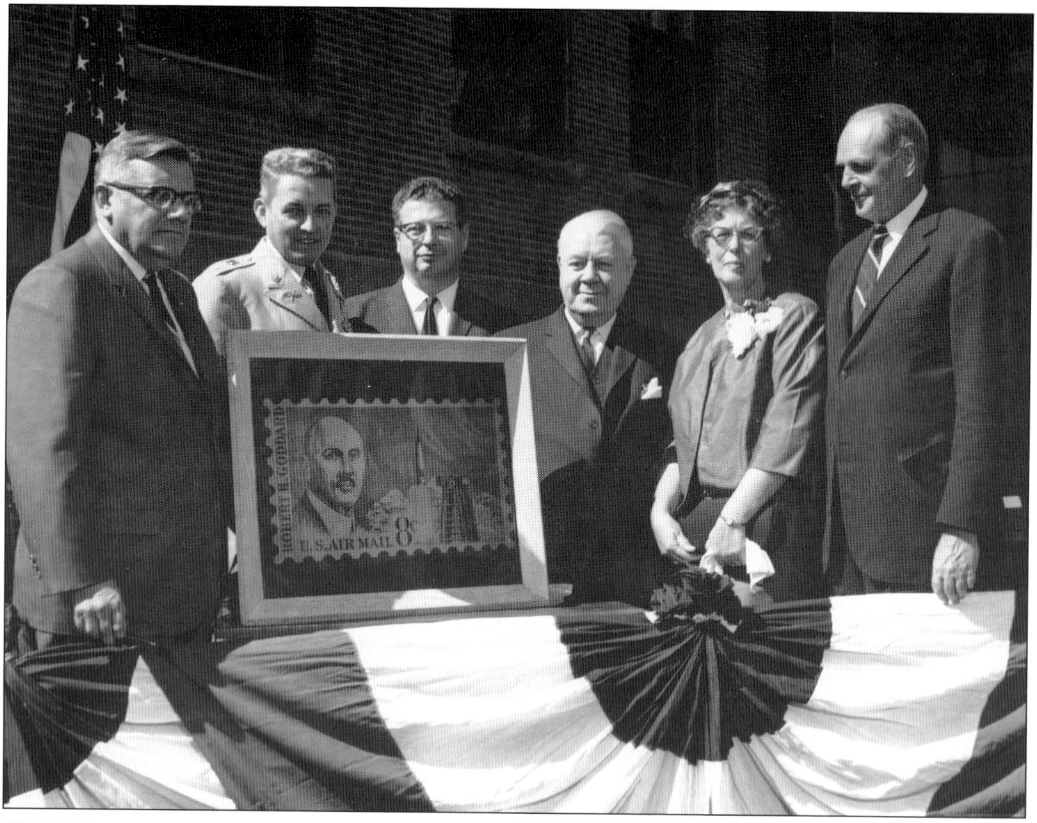

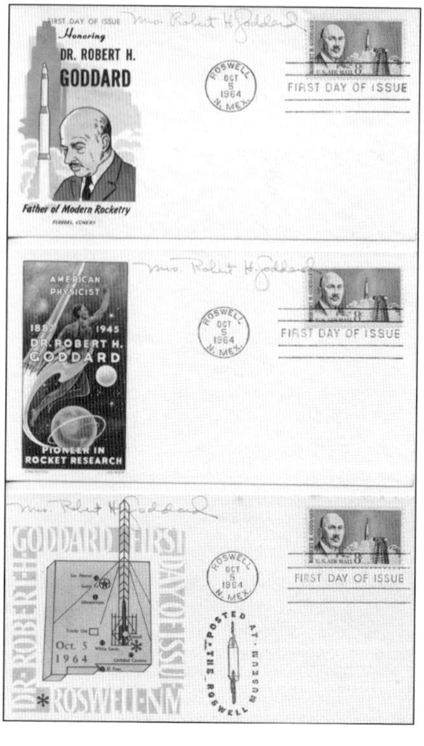

In 1964, the US Postal Service released an 8¢ stamp honoring Robert Goddard. Esther joined university and Post Office officials in unveiling the stamp in a ceremony held outside Clark University's physics building. Appropriately designated for airmail, the stamp was printed in yellow, red, and blue and featured a depiction of Robert next to a rocket being launched. The set of first day covers—envelopes with a new stamp postmarked on its first day of issue—for the stamp (left) features different caches depicting Robert Goddard and his research. (Both, Kevin Schindler personal collection.)

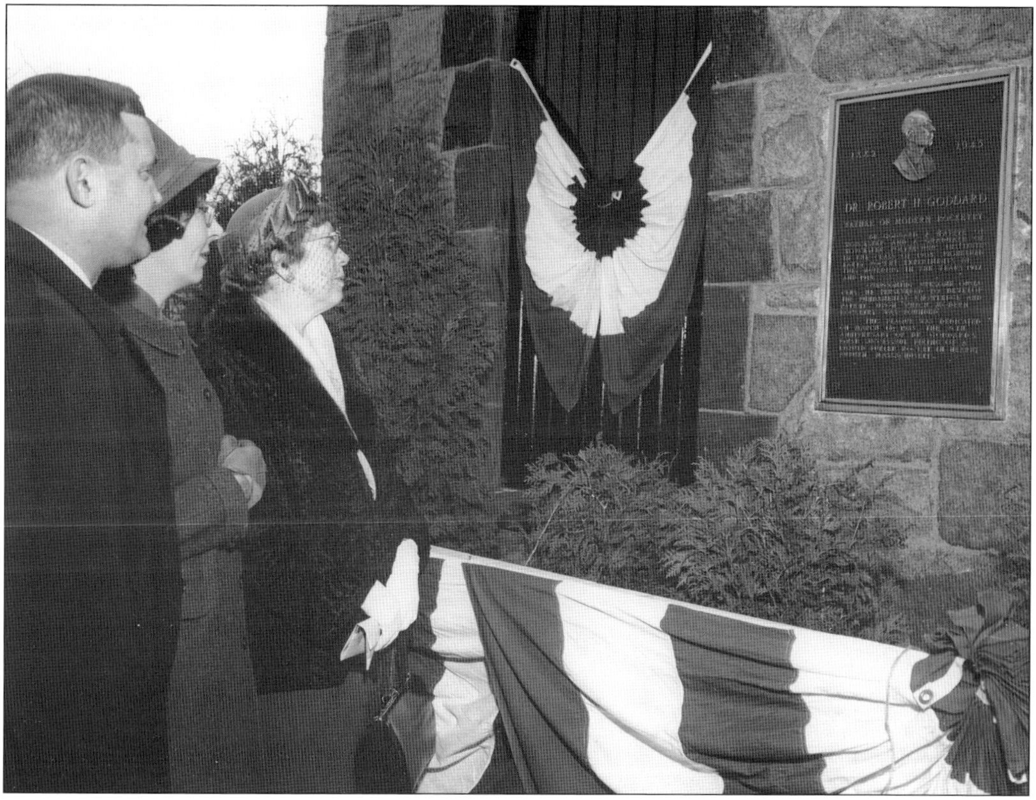

Worcester Polytechnic Institute celebrated Robert Goddard Day on March 16, 1965, by unveiling a plaque recognizing Goddard that was mounted onto the exterior north wall of Skull Tomb, a facility where he had conducted experiments a half-century earlier. An excerpt of the plaque reads, "Overcoming awesome obstacles, he persevered to develop the fundamental knowledge and patents upon which modern rocketry was founded." As shown in this picture, Esther joined the ceremony.

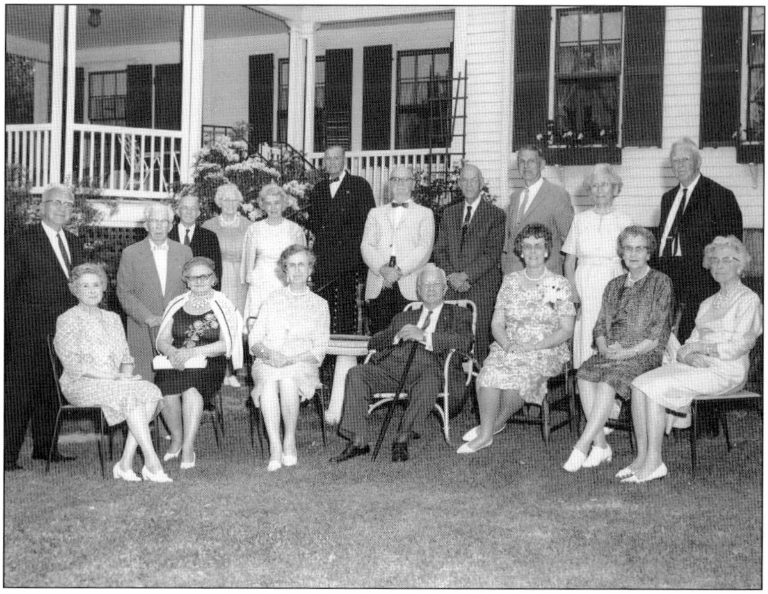

On June 11, 1965, Esther hosted a reception for members of WPI's 1908 graduating class—Robert's class. Though 57 years had passed and many, including Robert, had died, the gathering at the Goddard home offered a moment to reflect and celebrate. In the photograph taken that day, Esther sits third from the right in the first row.

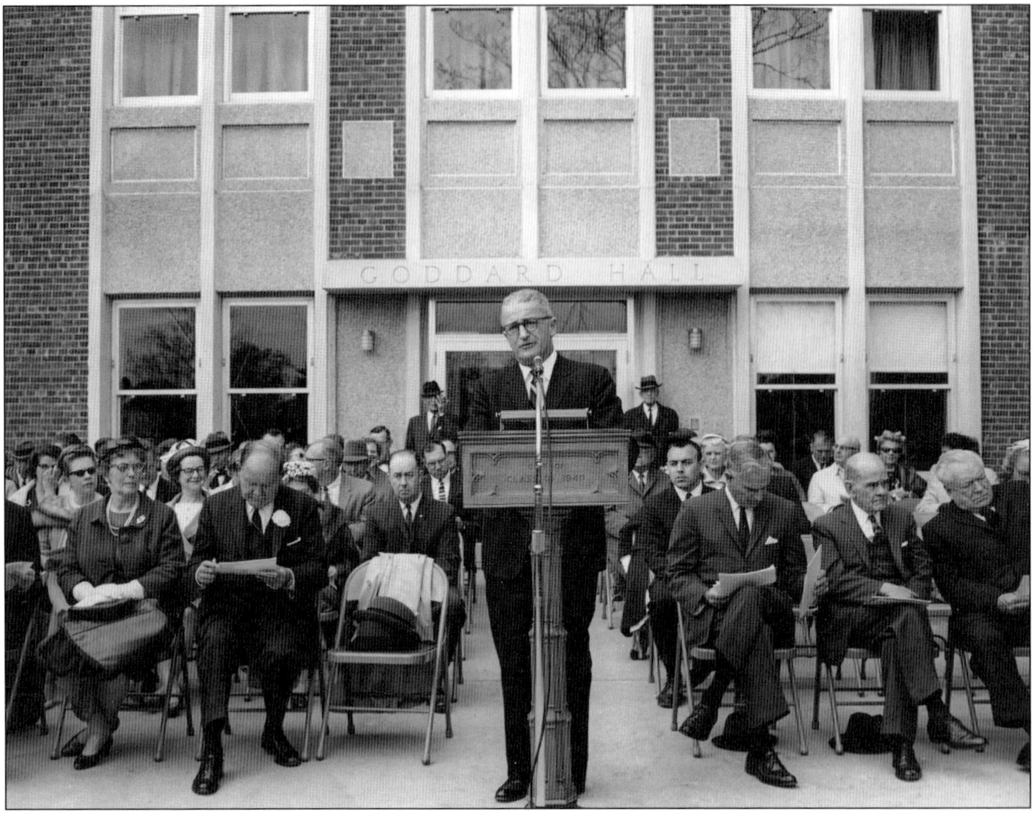

Another 1965 milestone celebrating Robert Goddard was WPI's opening of a new building—named Goddard Hall—to house its chemistry, biochemistry, and chemical engineering departments. During the ceremony, Esther attended and is seen in this picture seated in the first row on the left, continuing her role as a gracious steward of Robert's memory.

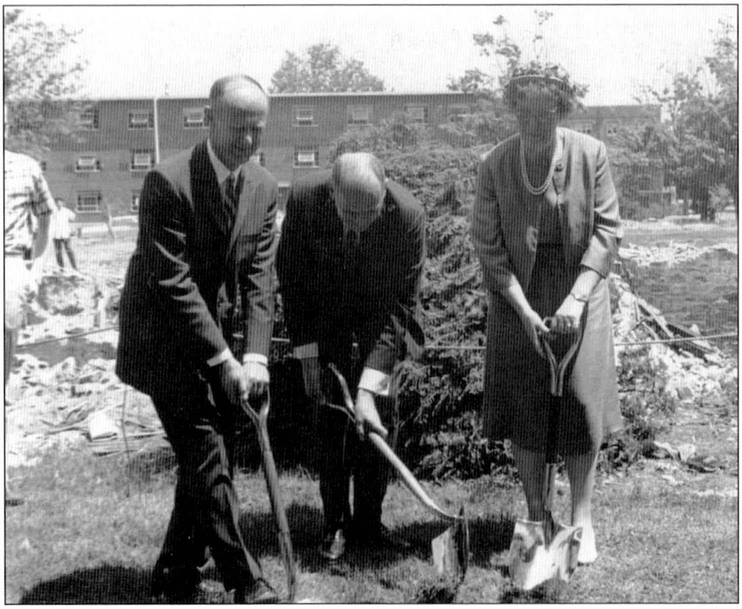

In December 1956, Clark University began planning for a new library that would eventually be named in honor of Robert Goddard. A decade later, in 1966, dignitaries gathered for the ceremonial ground-breaking. Esther Goddard was present to help turn the first shovels of dirt, marking a proud moment in preserving her husband's legacy at the university where he once taught.

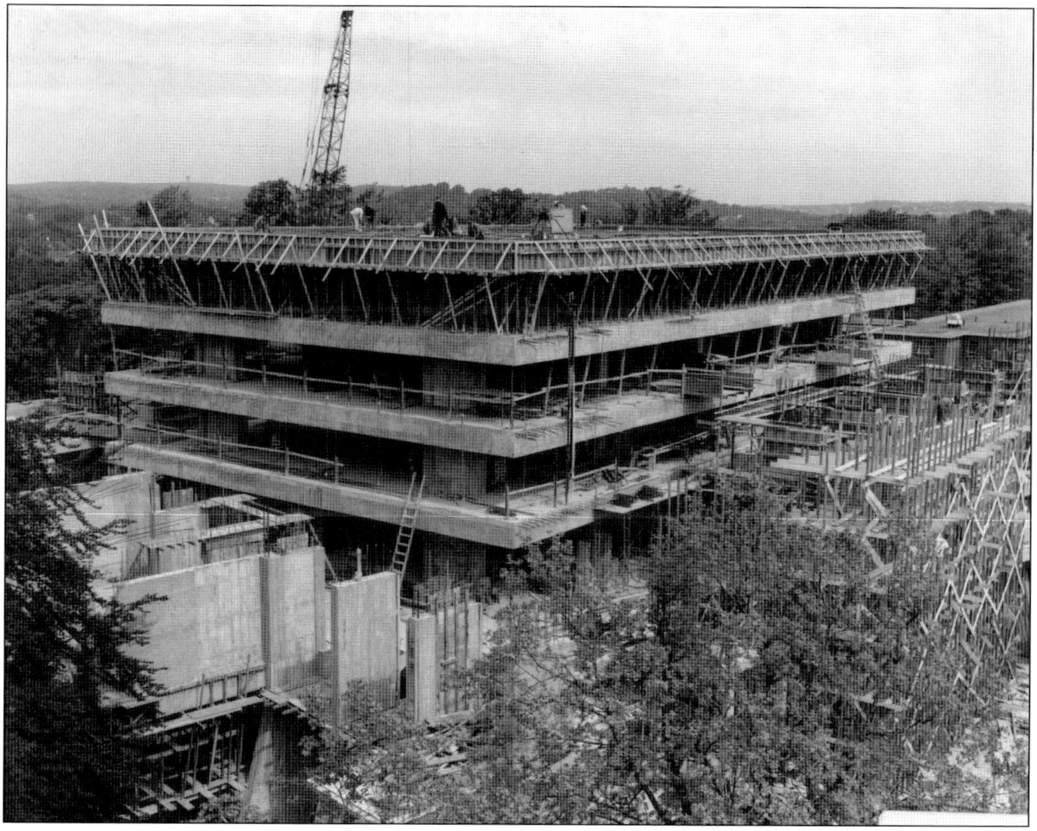

Construction of the Robert Hutchings Goddard Library at Clark University progressed from 1966 to 1969. Architect John Johanson of Connecticut designed the 134,000-square-foot facility in the brutalist style, characterized by the use of bare concrete, minimal decorative elements, and distinct geometrical shapes. As a symbol of innovation, the building reflects the spirit of its namesake.

A unique element of the Robert Hutchings Goddard Library is located near the entry to the special collections area. A large plaque on the floor marks the spot where a time capsule is buried. The capsule was assembled and dedicated in 1966—five months after ground-breaking for the library—in a ceremony attended by dignitaries that included US vice president Hubert Humphrey.

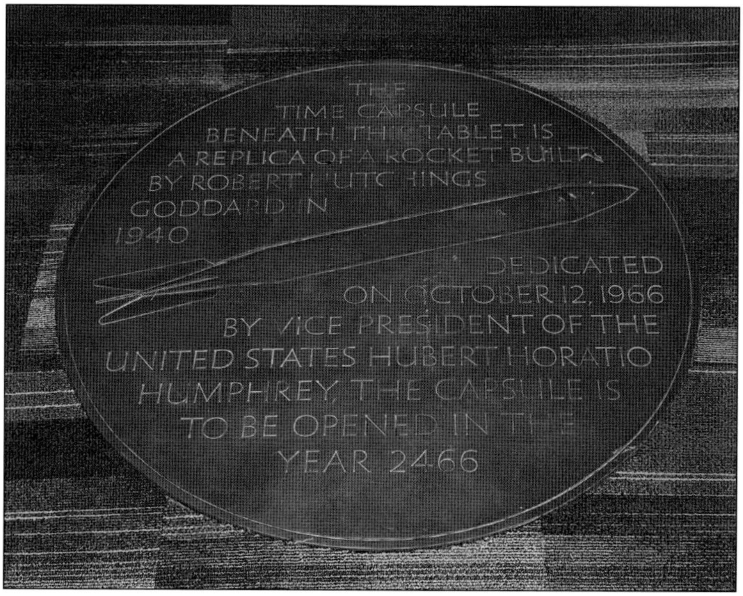

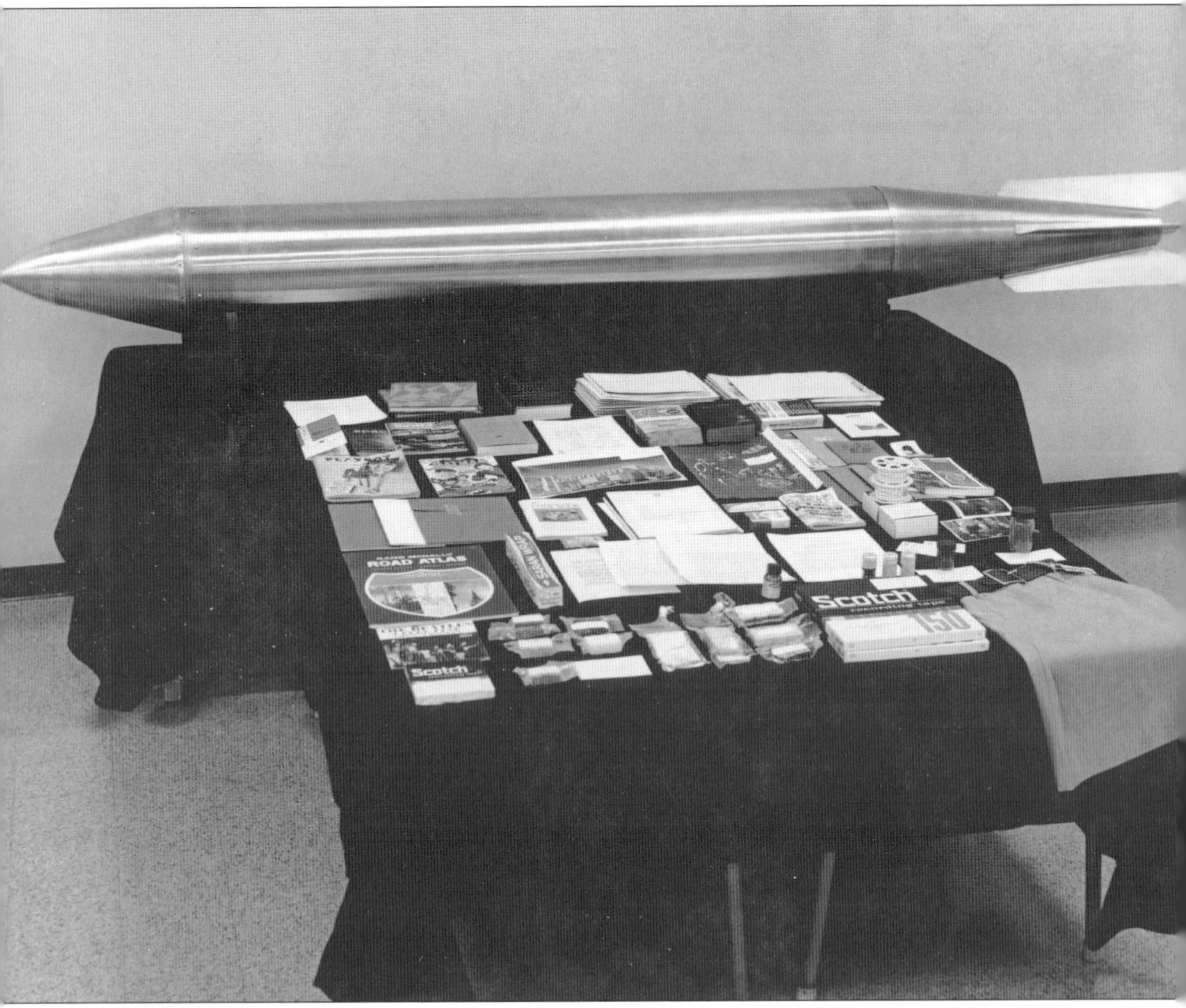

The Goddard Library time capsule consists of a 10-foot-long, 300-pound stainless steel and aluminum replica of one of Robert Goddard's rockets. A 55-member committee selected 90 items to be included in the capsule, reflecting four themes: Goddard's legacy, the Space Age, Clark University, and contemporary life. Among the contents are Goddard's honorary degree from Clark University, personal correspondence with science fiction writer H.G. Wells, and a vial of soil from Goddard's New Mexico rocket testing site. To capture the spirit of the Space Age, the capsule includes a film of astronaut Ed White's historic first American spacewalk and a retrorocket from an Apollo spacecraft. A miniskirt, a Beatles record, a *Batman* comic book, and tranquilizer pills represent the 1960s and contemporary life. Clark University's contents include a photograph of Sigmund Freud and the will of Jonas Clark, the university's founder. This eclectic mix offers future generations a snapshot of Robert Goddard and his impact on history, from Worcester to outer space, as well as a feel for mid-20th-century life.

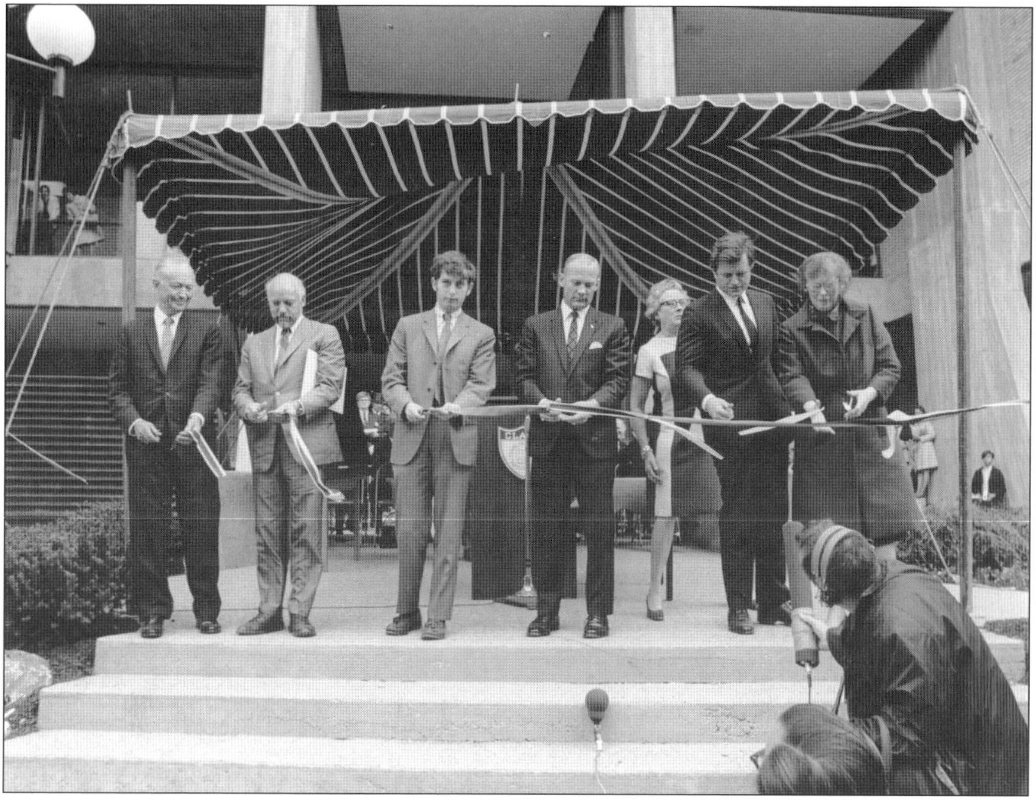

On May 19, 1969, Clark University honored one of its most distinguished alumni with the dedication of the Robert Hutchings Goddard Library. Dignitaries participating in the ribbon-cutting ceremony included, from left to right, J. Leland Atwood, president of North American Rockwell Corp.; John Johansen, the library's architect; Michael Feldman, president of the student council; Buzz Aldrin, Apollo 11 astronaut; Edward Kennedy, US senator; and Esther Goddard, Robert's widow.

The dedication of the Robert Hutchings Goddard Library marked a symbolic link between Goddard's pioneering work and the fulfillment of his dream of space travel in the form of the upcoming Apollo 11 moon landing. Buzz Aldrin (right) attended the ceremony alongside Goddard's mentor and former Smithsonian secretary Charles Abbot (left) and Aldrin's father, Edwin Aldrin Sr. (center), who studied under Goddard at Clark and, like Buzz, became an aviator.

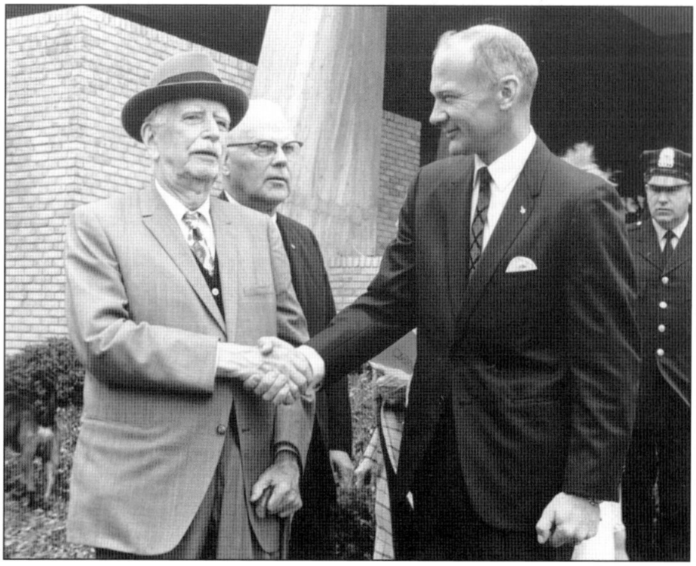

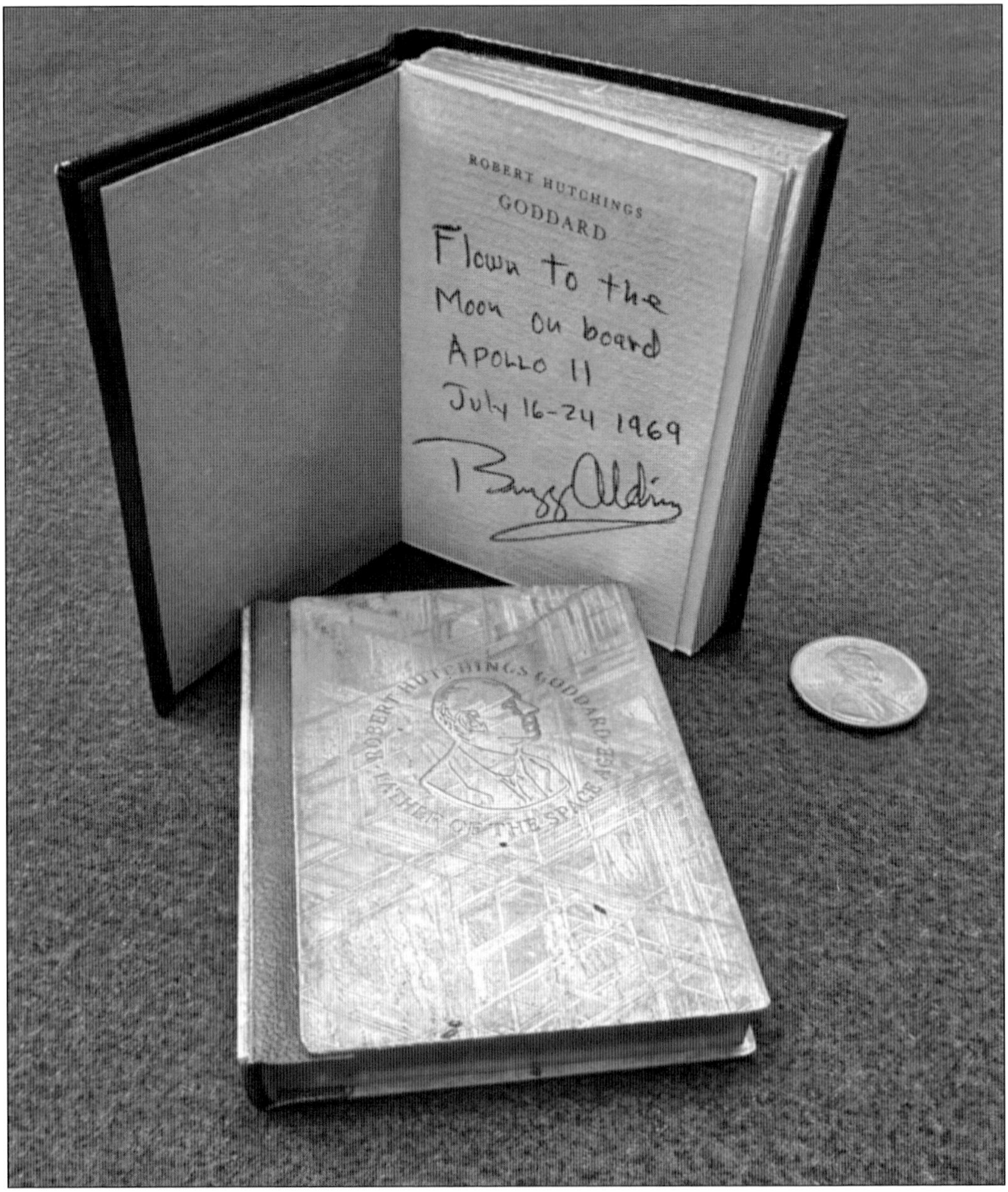

Just two months after the dedication of the Robert Hutchings Goddard Library at Clark University, astronaut Buzz Aldrin carried a miniature autobiography written by Robert Goddard to the Moon aboard Apollo 11. The book, *Robert Hutchings Goddard–Father of the Space Age*, was published by Worcester native Achille St. Onge, who was known for his exquisitely crafted miniature volumes. Bound in blue leather with gold-leaf edges, it became the first book ever flown to the lunar surface. Aldrin gifted the book to Robert's widow, Esther, who later donated it to the Goddard Library. Years later, Swiss book collector Arno Gschwendtner donated another copy of the autobiography to the library. This edition was bound with a sliced fragment of a meteorite, symbolizing the cosmic reach of Robert's legacy. Together, these artifacts serve as powerful tributes to Robert Goddard's enduring influence on space exploration and to the cycle of inspiration that continues to propel humanity skyward.

Another tribute to Goddard stands on the approach to the Robert Hutchings Goddard Library at Clark University. Embedded in a block of stone, a plaque features a quote from Charles Lindbergh, the legendary aviator and longtime supporter of Goddard's work: "We have touched the stars he reached for." (Kevin Schindler personal collection.)

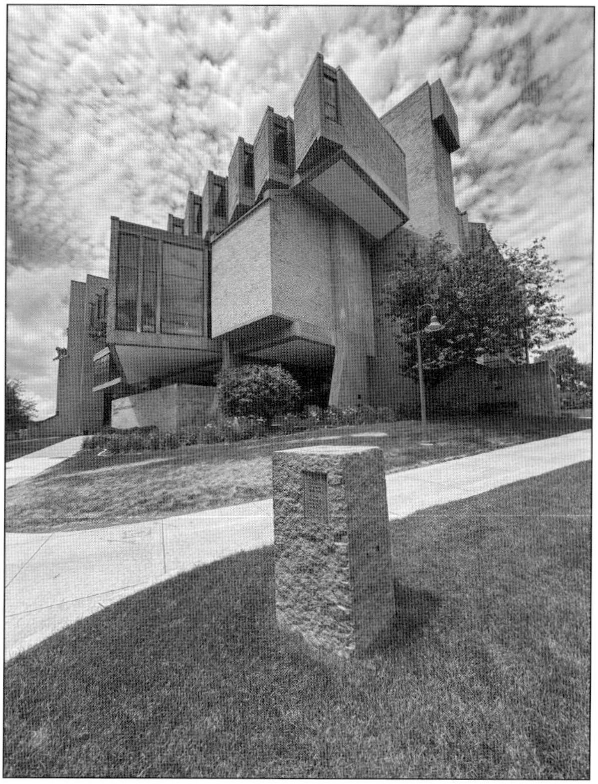

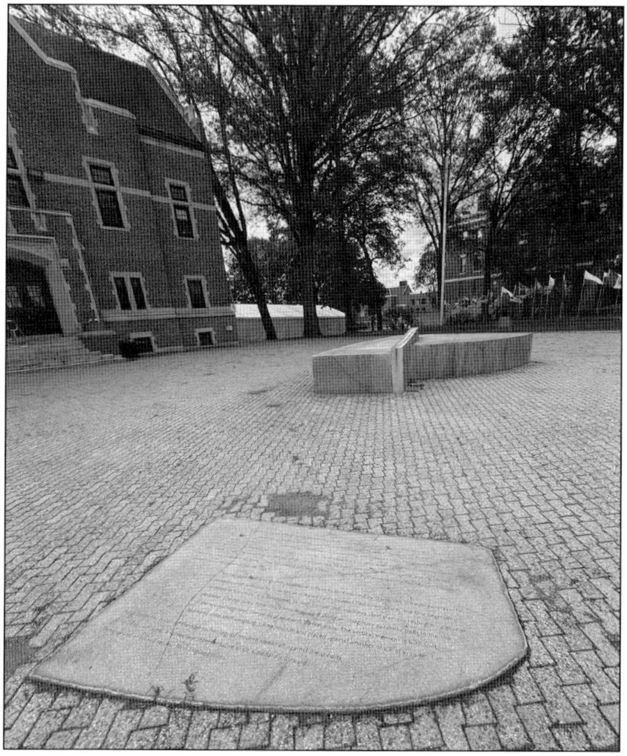

In 1978, Clark University dedicated the Goddard Memorial sculpture in front of Atwood Hall, where Robert Goddard had established the radio station WCUW in 1920. The two-part structure, crafted from both granite and steel, was designed by abstract artist and sculptor Dave von Schlegell to represent the parabolic trajectory of a rocket leaving Earth. (Kevin Schindler personal collection.)

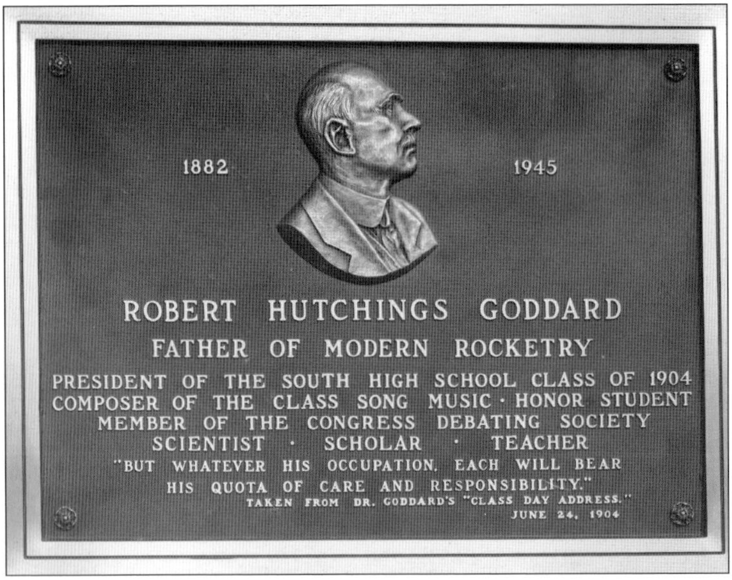

Robert Goddard's high school alma mater, South High, also recognized its favorite son with a plaque, as seen in this 1969 picture. South High eventually outgrew the building it had occupied during Goddard's time at 14 Richards Street and relocated to 170 Apricot Street in 1978. The old South High building briefly served as a middle school before becoming home to the Goddard School of Science and Technology.

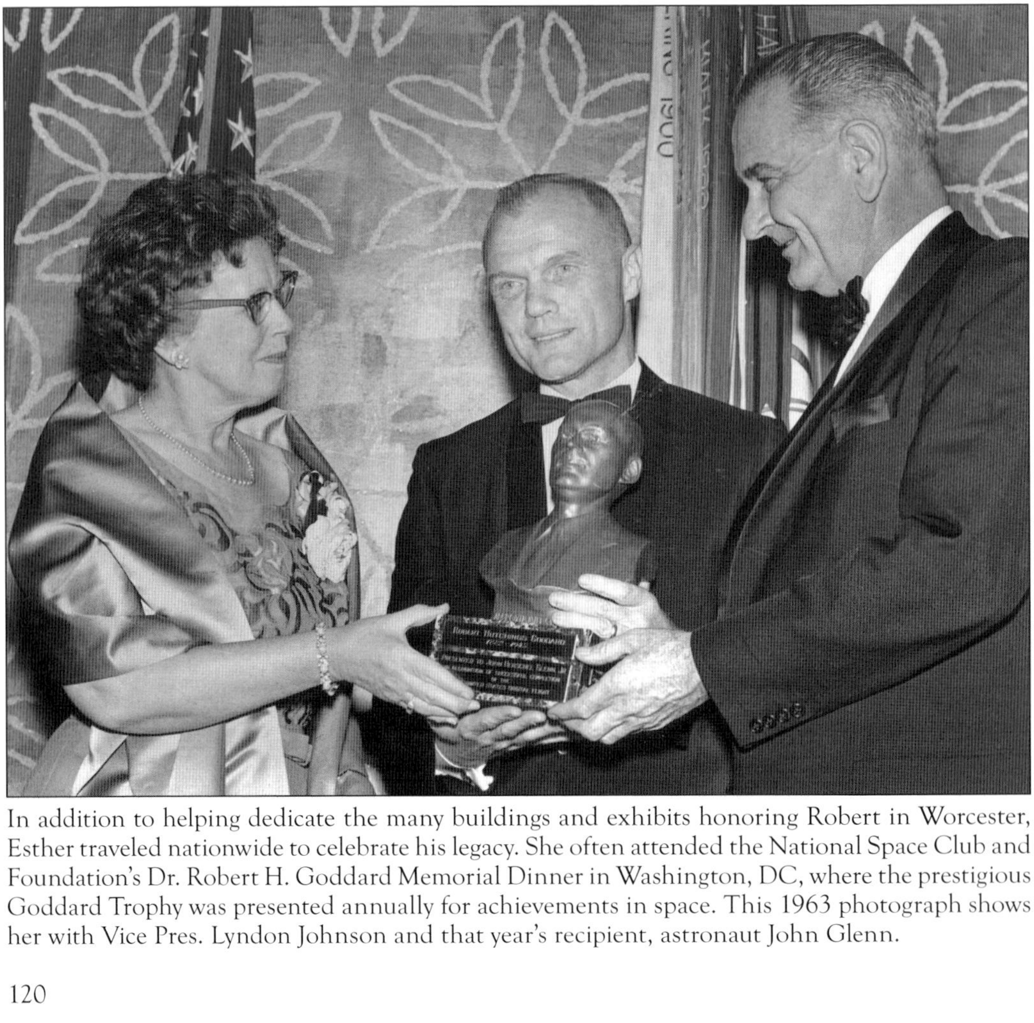

In addition to helping dedicate the many buildings and exhibits honoring Robert in Worcester, Esther traveled nationwide to celebrate his legacy. She often attended the National Space Club and Foundation's Dr. Robert H. Goddard Memorial Dinner in Washington, DC, where the prestigious Goddard Trophy was presented annually for achievements in space. This 1963 photograph shows her with Vice Pres. Lyndon Johnson and that year's recipient, astronaut John Glenn.

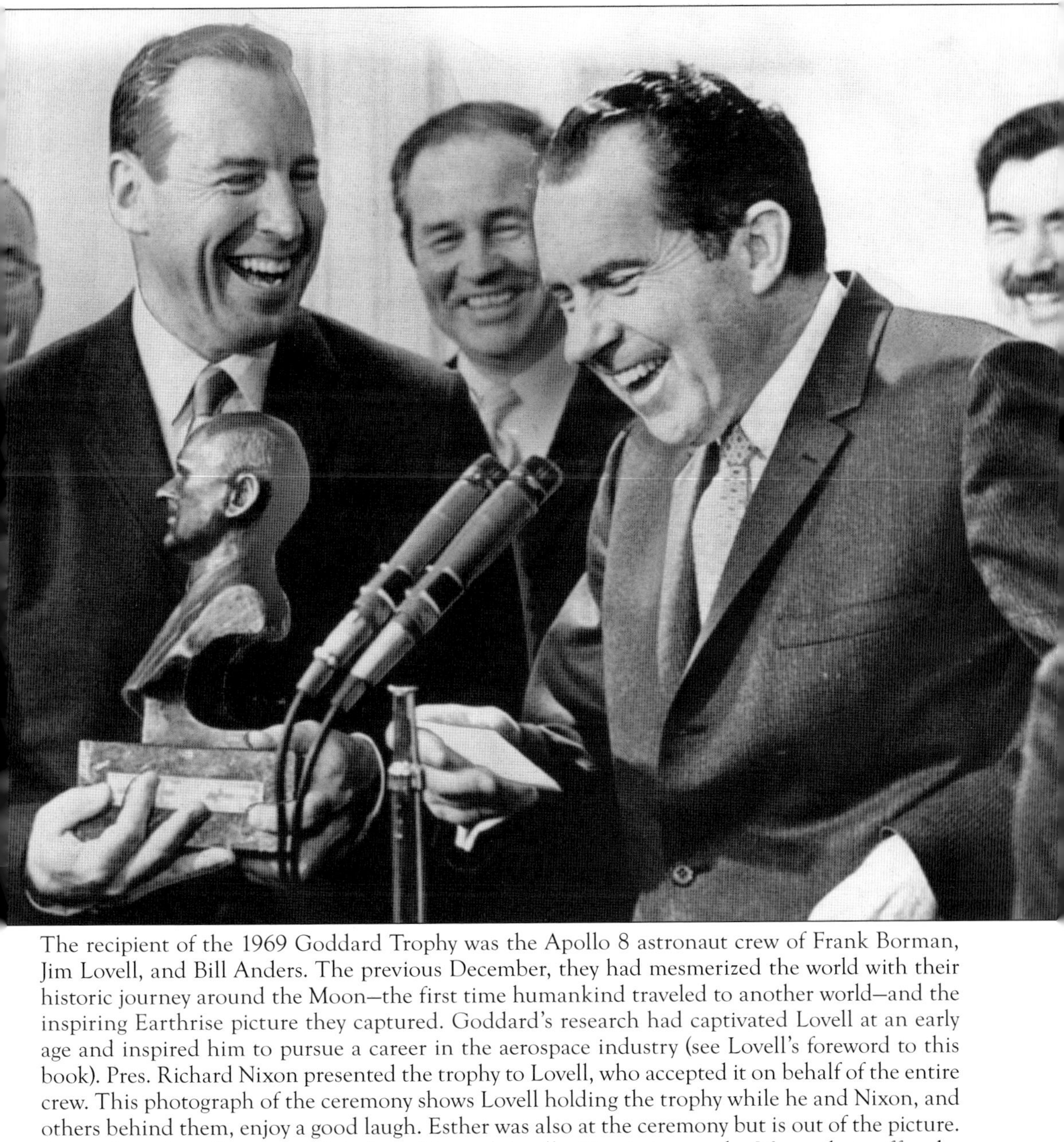

The recipient of the 1969 Goddard Trophy was the Apollo 8 astronaut crew of Frank Borman, Jim Lovell, and Bill Anders. The previous December, they had mesmerized the world with their historic journey around the Moon—the first time humankind traveled to another world—and the inspiring Earthrise picture they captured. Goddard's research had captivated Lovell at an early age and inspired him to pursue a career in the aerospace industry (see Lovell's foreword to this book). Pres. Richard Nixon presented the trophy to Lovell, who accepted it on behalf of the entire crew. This photograph of the ceremony shows Lovell holding the trophy while he and Nixon, and others behind them, enjoy a good laugh. Esther was also at the ceremony but is out of the picture. A year later, Lovell commanded the celebrated Apollo 13 mission to the Moon that suffered a critical explosion but nevertheless made it safely back to Earth. A year after that, in 1971, Lovell traveled to Worcester and joined Esther in dedicating the new Worcester Science Center. (Kevin Schindler personal collection.)

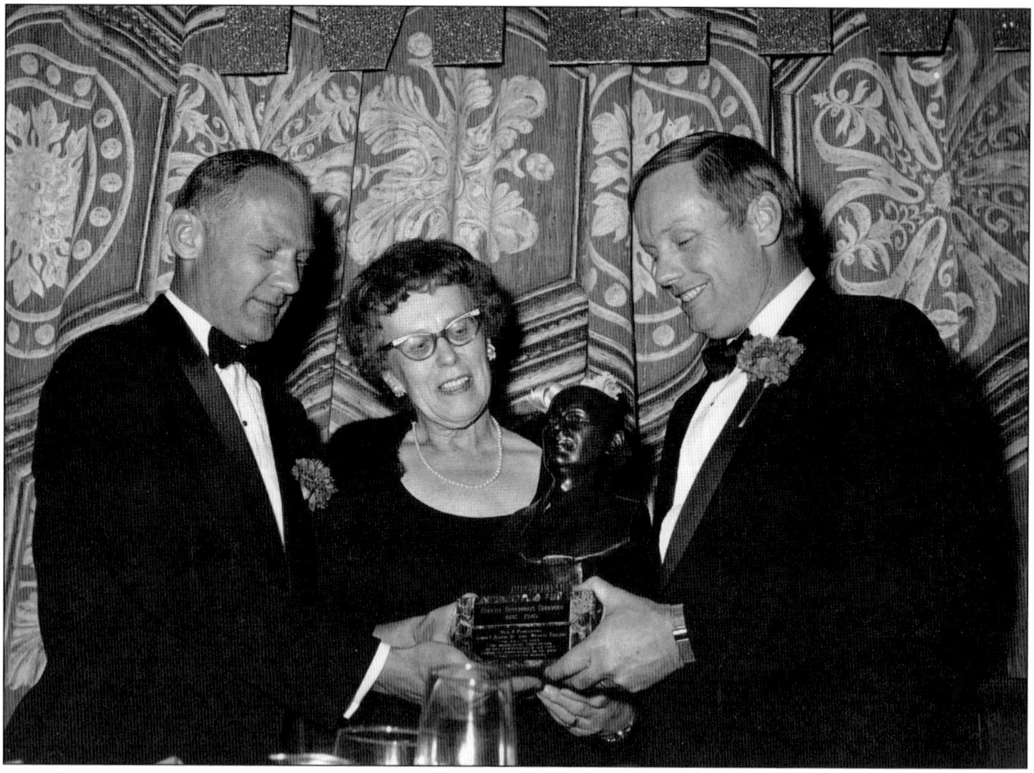

Esther returned to Washington, DC, in 1970 for the presentation of the Goddard Trophy to the crew of Apollo 11 in recognition of their historic landing on the Moon the previous year. The ceremony marked a full circle experience of sorts—Robert Goddard had taught physics to Buzz Aldrin's father, Edwin Sr., who then became an accomplished aviator. Edwin's son, Buzz, followed in his father's footsteps and became an aviator on his way to helping fulfill Robert's dream of space travel to other celestial bodies. Esther Goddard stands here with Buzz Aldrin (left) and Apollo 11 crewmate Neil Armstrong.

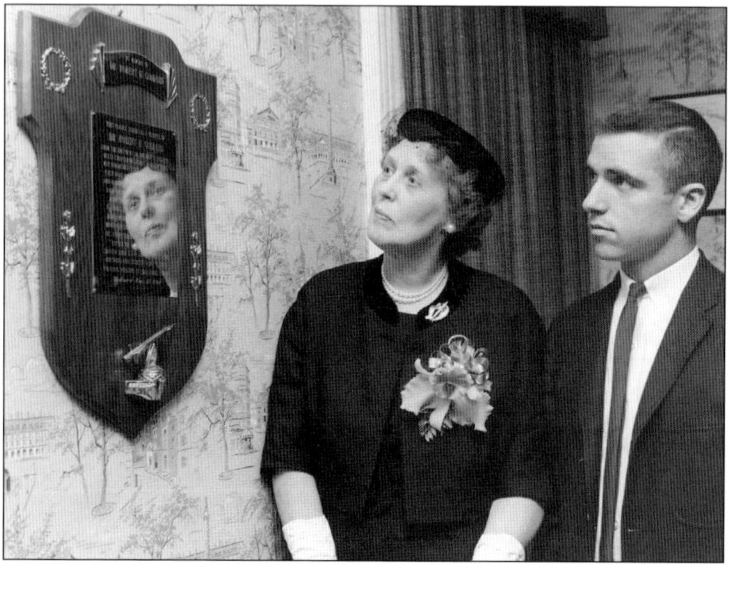

In 1972, members of the Sigma Alpha Epsilon fraternity at WPI unveiled in their chapter house a commemorative plaque honoring Robert Goddard, a member of the fraternity during his undergraduate years. Esther joined fraternity brothers and invited guests in honoring her late husband's legacy. The plaque recognized Goddard's pioneering work in rocketry and his lasting connection to WPI.

On May 12, 2006, the Goddard Memorial Association dedicated the Robert Goddard Monument at the corner of Goddard Memorial Drive and Apricot Street in Worcester. The 12-foot-wide monument features eight panels that highlight the life and work of Robert and Esther Goddard. These panels encircle a 10-foot-high, stylized stainless steel rocket, symbolizing Robert's contributions to the exploration of space. (Kevin Schindler personal collection.)

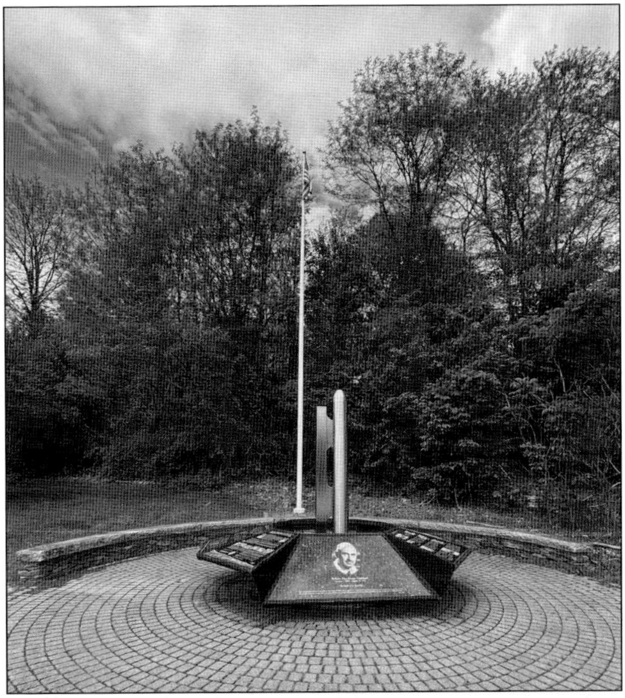

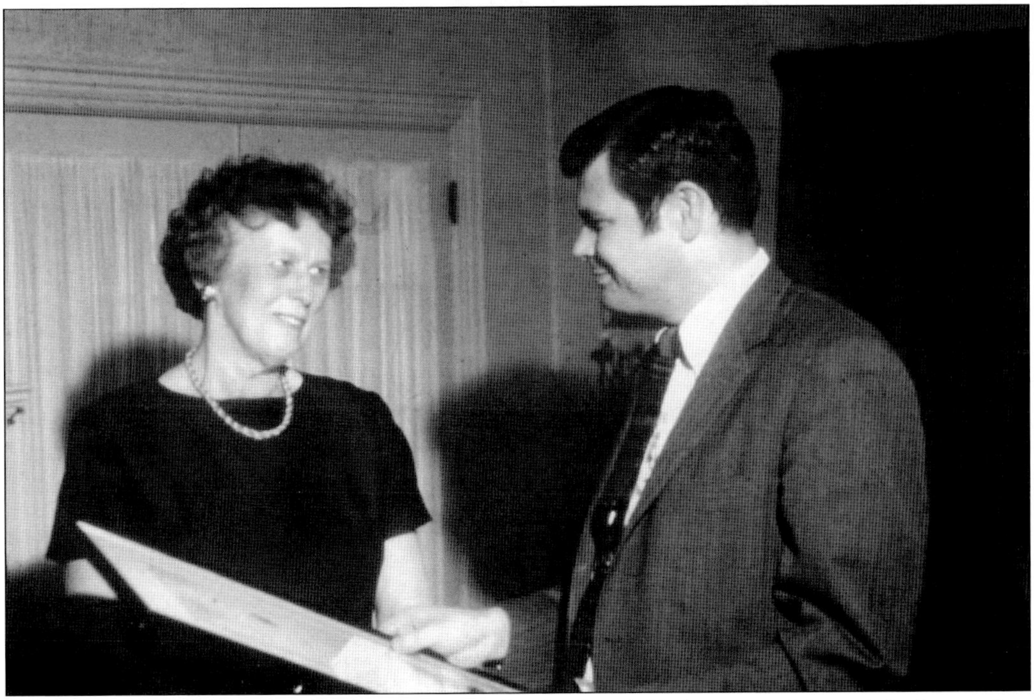

Beyond attending ceremonial events honoring her husband, Esther was celebrated for her tireless efforts to preserve his legacy. In this c. 1970 photograph, Rochester Museum of Science executive director Ian McLennan presents Esther with a certificate recognizing her as a fellow of the museum. This recognition highlights the impact of her dedication and the importance of her role in safeguarding her husband's legacy for future generations. (Ian McLennan.)

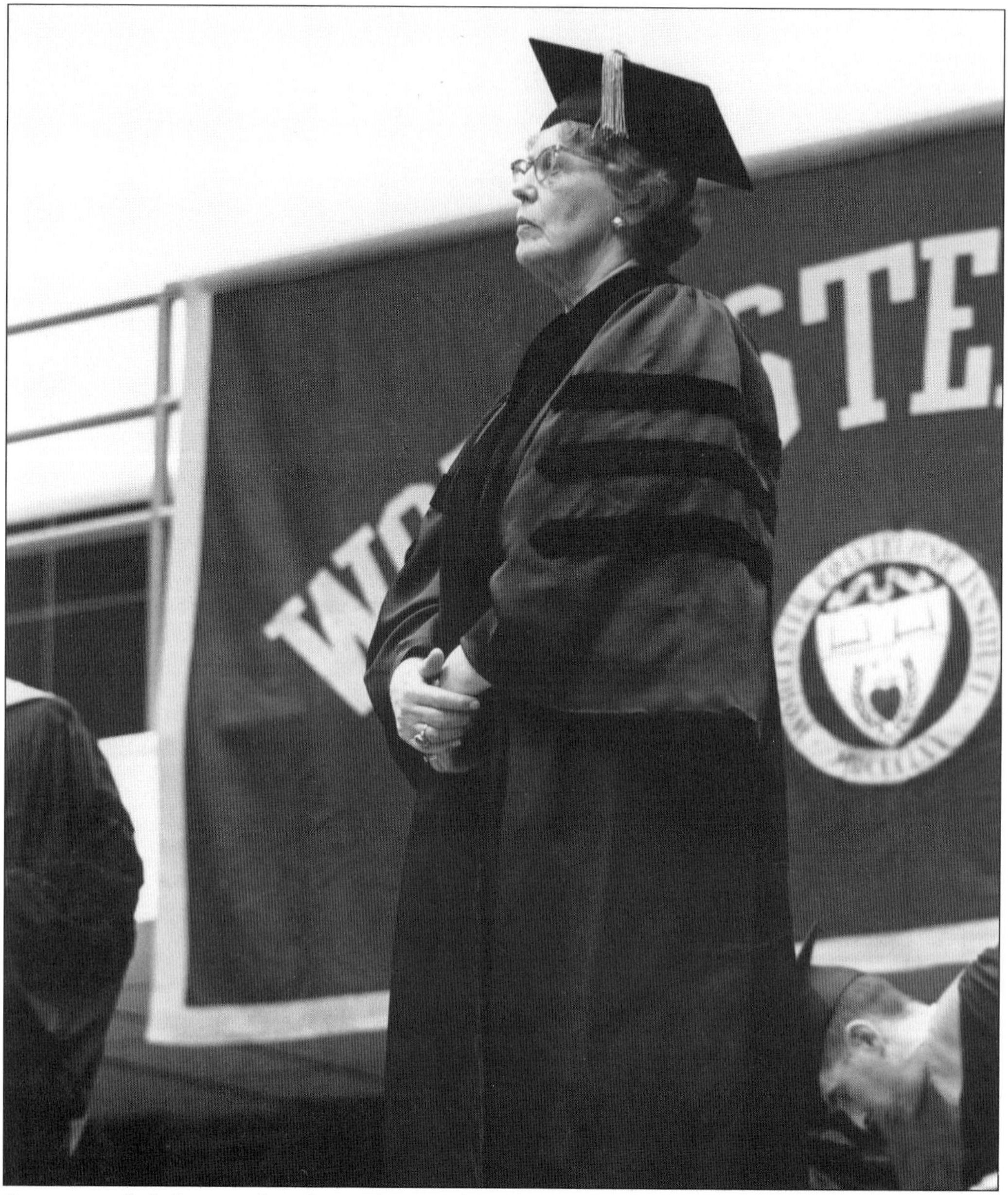

One cannot help but wonder what Robert Goddard's career and legacy would have been without the unflinching support of Esther. Robert earned his doctorate in 1911 and conducted his pioneering research until 1945, a span of 35 years. Esther labored to preserve Robert's legacy from the time of his death in 1945 essentially until her own death in 1982, a total of 38 years. This means she spent more time preserving and promoting his legacy than his scientific career lasted. And that does not include the years she supported him while he was alive, photographing and filming launches, transcribing notes, and managing logistics. For her part, Esther started attending college when she was younger but left school when she married Robert. While they lived in Maryland from 1942 to 1945, she resumed her education and earned a bachelor of arts degree, with honors, from Johns Hopkins University. Years later, in 1969, WPI recognized her with an honorary doctor of science award, as depicted in this picture.

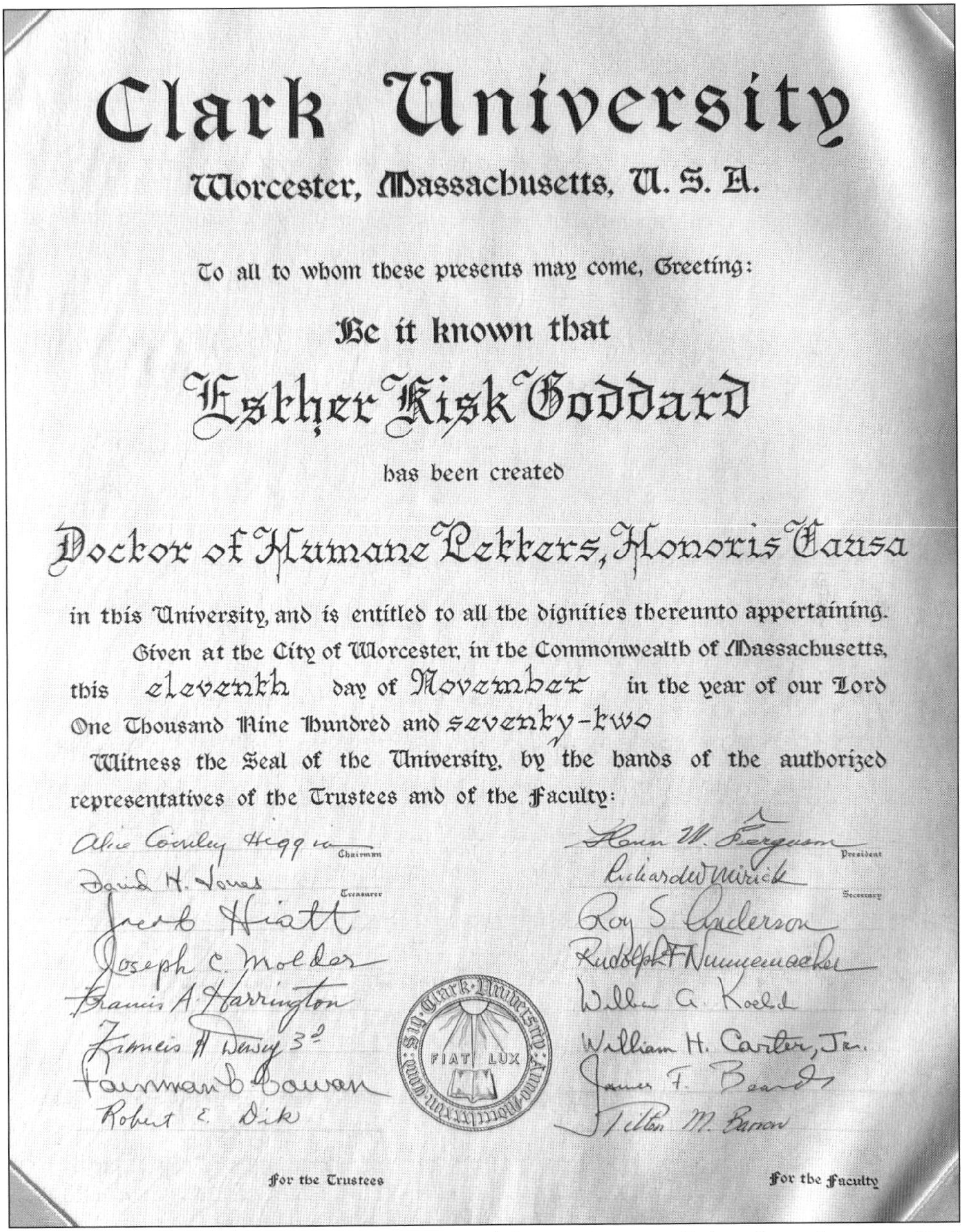

Clark University

Worcester, Massachusetts, U. S. A.

To all to whom these presents may come, Greeting:

Be it known that

Esther Kisk Goddard

has been created

Doctor of Humane Letters, Honoris Causa

in this University, and is entitled to all the dignities thereunto appertaining.

Given at the City of Worcester, in the Commonwealth of Massachusetts, this eleventh day of November in the year of our Lord One Thousand Nine Hundred and seventy-two

Witness the Seal of the University, by the hands of the authorized representatives of the Trustees and of the Faculty:

Chairman President

Treasurer Secretary

FIAT LUX

For the Trustees For the Faculty

In addition to the honorary degree she received from Worcester Polytechnic Institute, Esther Goddard was also honored by Clark University. In 1972, Clark presented her with a doctor of humane letters, recognizing her tireless efforts to preserve and promote Robert's legacy. Esther had played a vital role in curating Robert's influential contributions to science, donating a vast collection of his documents, artifacts, photographs, notebooks, and other items to Clark's archives and special collections, housed in the Robert Hutchings Goddard Library. The materials included the rocket support base from an early flight, rare manuscripts, technical drawings, and the miniature autobiography that Buzz Aldrin took to the Moon during the Apollo 11 mission. Esther's donations ensured that future generations could study the groundbreaking work of one of America's foremost aerospace pioneers. (WPI, George C. Gordon Library.)

Bibliography

Beals, Everett. "Outta this World: Alumnus Donates Miniature Book Bound in Meteorite." *Scarlet*, January 30, 2024. thescarlet.org/19283/news/outta-this-world-alumnus-donates-miniature-book-bound-in-meteorite/.

Clary, Robert. *Rocket Man*. New York: Hyperion, 2004.

Goddard, Esther, and G. Edward Pendray. *The Papers of Robert H. Goddard*. 3 vols. New York: McGraw-Hill, 1970.

Goddard, Robert H. *Autobiographical Statement*. Robert Hutchings Goddard Library, Clark University, 1921.

———. *Diaries of Robert H. Goddard, 1898–1945*. Robert Hutchings Goddard Library, Clark University.

———. *Material for an Autobiography*. Robert Hutchings Goddard Library, Clark University, 1927, rev. 1933.

———. *A Method of Reaching Extreme Altitudes*, 1919. Smithsonian Miscellaneous Collections 71, no. 2. Smithsonian Institution.

Keough, Jim. "The Write Stuff: Robert Goddard Biography Was the First Book on the Moon." Clark University. September 15, 2011. www.clarku.edu/news/2011/09/15/the-write-stuff/.

Lehman, Milton. *This High Man: The Life of Robert H. Goddard*. New York: Farrar, Straus and Company, 1963.

Von Braun, Wernher, and Frederick I. Ordway. *History of Rocketry and Space Travel*. New York: Thomas Y. Crowell, 1966. Reprint 1969.

The Wonder Mission: Bringing the Universe to Main Street. thewondermission.org.

Yeulenski, Lois R. *Worcester*. Charleston, SC: Arcadia Publishing, 1994. Reprint 1998.

About the Wonder Mission

Bringing the Universe to Main Street

For more than a century, the United States stood at the forefront of science, technology, engineering, and space exploration. From the Wright brothers to the Apollo missions, American ingenuity has shaped the modern world. Yet, by the 2020s, the United States was losing its position of leadership in these areas. As of the writing of this book, more than 1.5 million STEM jobs remain unfilled, and many individuals, young and old, feel disconnected from the remarkable achievements unfolding around them.

The Wonder Mission, a 501(c)(3) nonprofit organization, was founded to reverse this by launching a series of initiatives to help alter this troubling trend and inspire our next generation of those who dream big.

Dr. Robert Goddard's story perfectly illustrates the values needed to promote and encourage: vision, perseverance, ingenuity, and invention. The Robert Goddard Project is working to publicize and share Goddard's remarkable legacy and vision but also, just as importantly, to inspire the Goddards of today and tomorrow—the scientists, educators, students, engineers, innovators, and dreamers dedicated to expanding scientific knowledge for the benefit of humanity.

Other Wonder Mission initiatives include saving and restoring Goddard's home in Worcester, launching the National Space Trail (starting at Goddard's home), and encouraging others across the nation to leverage Goddard's amazing legacy for the greater good.

Through the publication of this book and a series of creative initiatives and outreach efforts, the Wonder Mission aims to spark imagination and remind the public that anything is possible. As Goddard once said, "The dream of yesterday is the hope of today and the reality of tomorrow."

To support the Wonder Mission in inspiring people of all ages to dream big, explore boldly, and help shape a future where science and wonder are part of everyday life, please visit thewondermission.org.